D1104535

STATISTICAL
PROCESS CONTROL
IN MANUFACTURING

QUALITY AND RELIABILITY

A Series Edited by

Edward G. Schilling

Center for Quality and Applied Statistics
Rochester Institute of Technology
Rochester, New York

STATISTICAL PROCESS CONTROL IN MANUFACTURING

edited by

J. BERT KEATS
DOUGLAS C. MONTGOMERY
Arizona State University
Tempe, Arizona

MARCEL DEKKER, INC.   **NEW YORK • BASEL • HONG KONG**
ASQC QUALITY PRESS   **MILWAUKEE**

Library of Congress Cataloging-in-Publication Data

Statistical process control in manufacturing / edited by J. Bert
   Keats, Douglas C. Montgomery.
     p. cm. --- (Quality and reliability ; 23)
    Includes bibliographical references and index.
    ISBN 0–8247–8467–7 (alk. paper)
    1. Process control---Statistical methods. 2. Quality control---
Statistical methods. I. Keats, J. Bert.
II. Montgomery, Douglas C. III. Series.
TS156.8.S753   1991
658.5----dc20   90--25092
                                          CIP

This book is printed on acid-free paper.

MARCEL DEKKER, INC.
270 Madison Avenue, New York, New York 10016

ASQC QUALITY PRESS
310 West Wisconsin Avenue, Milwaukee, Wisconsin 53203

Current printing (last digit):
10 9 8 7 6 5 4 3 2 1

PRINTED IN THE UNITED STATES OF AMERICA

ABOUT THE SERIES

The genesis of modern methods of quality and reliability will be found in a simple memo dated May 16, 1924, in which Walter A. Shewhart proposed the control chart for the analysis of inspection data. This led to a broadening of the concept of inspection from emphasis on detection and correction of defective material to control of quality through analysis and prevention of quality problems. Subsequent concern for product performance in the hands of the user stimulated development of the systems and techniques of reliability. Emphasis on the consumer as the ultimate judge of quality serves as the catalyst to bring about the integration of the methodology of quality with that of reliability. Thus, the innovations that came out of the control chart spawned a philosophy of control of quality and reliability that has come to include not only the methodology of the statistical sciences and engineering, but also the use of appropriate management methods together with various motivational procedures in a concerted effort dedicated to quality improvement.

This series is intended to provide a vehicle to foster interaction of the elements of the modern approach to quality, including statistical applications, quality and reliability engineering, management, and motivational aspects. It is a forum in which the subject matter of these various areas can be brought together to allow for effective integration of appropriate techniques. This will promote the true benefit of each, which can be achieved only through their interaction. In this sense, the whole of quality and reliability is greater than the sum of its parts, as each element augments the others.

The contributors to this series have been encouraged to discuss fundamental concepts as well as methodology, technology, and procedures at the leading edge of the discipline. Thus, new concepts are placed in proper perspective in these evolving disciplines. The series is intended for those in manufacturing, engineering, and marketing and management, as well as the consuming public, all of whom have an interest and stake in the improvement and maintenance of quality and reliability in the products and services that are the lifeblood of the economic system.

The modern approach to quality and reliability concerns excellence: excellence when the product is designed, excellence when the product is made, excellence as the product is used, and excellence throughout its lifetime. But excellence does not result without effort, and products and services of superior quality and reliability require an appropriate combination of statistical, engineering, management, and motivational effort. This effort can be directed for maximum benefit only in light of timely knowledge of approaches and methods that have been developed and are available in these areas of expertise. Within the volumes of this series, the reader will find the means to create, control, correct, and improve quality and reliability in ways that are cost effective, that enhance productivity, and that create a motivational atmosphere that is harmonious and constructive. It is dedicated to that end and to the readers whose study of quality and reliability will lead to greater understanding of their products, their processes, their workplaces, and themselves.

Edward G. Schilling

PREFACE

The past decade was characterized by a resurgence in statistical process control techniques and in the use of statistical design of experiments. While some have argued that with proper use of statistical design in the early stages of product and process development, there will be little need for the use of statistical process control, it is the opinion of the authors that both approaches are necessary. Indeed, the most dramatic improvements in variability reduction can be achieved through the use of statistical design of experiments in the development or re-design phases. However, it is our belief that statistical process control will continue to be a critical part of any company's quality activities. Many industrial processes are characterized by dynamic behavior of the process mean, correlated measures and/or the need for rapid, adaptive control. We are aware of several U.S. companies who have practiced statistical design of experiments for 20-30 years and yet continue to see a crucial need for statistical process control in their factories. Statistical process control will continue to be a viable tool in any continuous improvement program.

This book presents reports of recent developments and applications in both arenas. Most of the topics represent innovative approaches primarily intended for discrete manufacturing processes. Process control techniques in these chapters focus on dynamical systems and the need for control in real-time. Included in other chapters are design of experiment techniques and applications as well as other process analysis and improvement techniques.

The book is a blend of theoretical and applied topics. Eight of the chapters present methodology supported with use of industrial data. In our opinion, many of the topics in this volume represent new directions in quality technology.

Keeping pace with automated manufacturing has many implications--statistical process control procedures are needed for both high volume operations associated with transfer lines and small batch processes linked to flexible manufacturing systems and group technology. Sophisticated data acquisition systems are providing data at rates which were unheard of a few years ago. Process control engineers are being challenged to provide control at or near real time. Statistical control applications which are tied to static techniques will be of little value. What is needed are procedures for handling dynamic data (which the mean is constantly changing, although not necessarily so that the process is non-stationary), procedures for handling data which are not independent (characteristic of consecutive observations) and procedures which can make predictions about the next observation so that control may be applied proactively.

This volume was compiled with the notion of providing methodologies which would assist in "keeping pace". It includes other relevant procedures as well. It is our hope that quality practitioners will make use of what is being suggested if they find it appropriate in their regions of interest.

J. Bert Keats
Douglas C. Montgomery

CONTENTS

**Section   II   Experimental   Design
             for Product and Process Development**

Section   III   Process Analysis and Improvement

Contents

CONTRIBUTORS

Robert V. Baxley, Jr. Fellow, Nylon Process Technology Department, Monsanto Chemical Company, Pensacola, Florida

C-S. Cheng, Ph.D.* Graduate Student, Department of Industrial and Management Systems Engineering, Arizona State University, Tempe, Arizona

Yen Chu Member of the Technical Staff, Central Engineering Laboratories, FMC Corporation, Santa Clara, California.

Ken Chung, Ph.D. Member of the Technical Staff, Central Engineering Laboratories, FMC Corporation, Santa Clara, California.

John R. Davis, P.E. Senior Control Engineer, Profimatics, Inc., Thousand Oaks, California.

Frederick W. Faltin Statistician, Management Science and Statistics Program, GE Corporate Research and Development Center, Schenectady, New York.

Carol J. Feltz, Ph.D. Statistician, AT&T Technologies, Princeton, New Jersey.

---

*Present affiliation: Associate Professor, Department of Industrial Engineering, Yuan-Tze Memorial College of Engineering, Taoyuan Shian, Taiwan, Republic of China.

John J. Flaig*   Quality Engineer, Apple Computer, Inc., Cupertino, California.

Mikel J. Harry†   Chief Statistician and Member of the Technical Staff, Group Operations, Government Electronics Group, Motorola, Inc., Scottsdale, Arizona.

Gregory Hitchings   Senior Process Engineer, Ultramar, Long Beach, California.

Norma Faris Hubele, Ph.D.   Associate Director, Statistical and Engineering Applications for Quality Laboratory, CIM Systems Research Center, College of Engineering and Applied Sciences, Arizona State University, Tempe, Arizona.

Catherine M. Jablonsky   Consultant in Quality Sciences and Staff Associate, Luftig and Associates, Inc., Farmington Hills, Michigan.

J. Edward Jackson, Ph.D. Statistical Consultant, Rochester, New York.

J. Bert Keats, Ph.D.   Director, Statistical and Engineering Applications for Quality Laboratory, CIM Systems Research Center, College of Engineering and Applied Sciences, Arizona State University, Tempe, Arizona.

George F. Koons‡ Quantitative Problem Solving Group, McDonnell Aircraft Company, McDonnell Douglas Corporation, St. Louis, Missouri.

M. H. Lill   Contract Project Manager, Corporate Technology Center, FMC Corporation, Santa Clara, California.

Jeffrey J. Luner   Technical Specialist, Quantitative Problem Solving Group, McDonnell Aircraft Company, McDonnell Douglas Corporation, St. Louis, Missouri.

Current affiliations:
*Apple Computer, Inc., Santa Clara, California.
†Six Sigma Research Institute, Motorola University, Rolling Meadows, Illinois.
‡Motorola Lighting, Inc., Buffalo Grove, Illinois.

Steven A. Melnyk, Ph.D. Associate Professor, Department of Management, Graduate School of Business Administration, Michigan State University, East Lansing, Michigan.

Douglas C. Montgomery, Ph.D. Professor, Department of Industrial and Management Systems Engineering, Arizona State University, Tempe, Arizona.

Hamid Shahriari Graduate Student, Department of Industrial and Management Systems Engineering, Arizona State University, Tempe, Arizona.

George W. Sturm, Ph.D. Associate Professor, Department of Mathematics, Grand Valley State University, Allendale, Michigan.

Lora Svoboda Research Associate, Statistical and Engineering Applications for Quality Laboratory, CIM Systems Research Center, College of Engineering and Applied Sciences, Arizona State University, Tempe, Arizona.

William T. Tucker, Ph.D. Statistician, Management Science and Statistics Program, GE Corporate Research and Development Center, Schenectady, New York.

James F. Wolter, Ph.D. Associate Professor, Department of Marketing, Seidman School of Business, Grand Valley State University, Allendale, Michigan.

Steven A. Yourstone, Ph.D. Assistant Professor, Department of Management Science and Information Services, Anderson Schools of Management, University of New Mexico, Albuquerque, New Mexico

Mona A. Yousry, Ph.D. Statistician, AT&T Technologies, Princeton, New Jersey.

1
INTRODUCTION

J. Bert Keats and
Douglas C. Montgomery
Arizona State University
Tempe, AZ 85287

Over half of the papers in this book are based on presentations made at the Third National Symposium, "Statistics in Design and Process Control: Keeping Pace with Automated Manufacturing" sponsored by Arizona State University and held in Tempe, Ariz., on November 14-16, 1988. The other papers were solicited by the editors to represent other relevant and recent developments.

This volume is presented in three sections:

Section I presents papers which focus on statistical process control applications.
Section II provides four papers relating to statistics and design, two of which are case studies.
Section III addresses process capability, analysis and improvement.

SECTION 1

Koons and Luner attack the problem of providing statistical process control in a low volume manufacturing environment through the use of deviations from nominal as the metric of merit and an emphasis on the process rather than the product. Their investigation then focuses on whether the process is operating consistently and capably. The manufacturing lot is selected as the rational subgroup. Koons

1

and Luner illustrate the use of their methodology with data from side milling operations in an aircraft manufacturing facility.

Also concerned with small volume manufacturing are Lill, Chu and Chung. However, their focus is in identifying and reducing variability of set-up operations. They assume that total variability is the sum of set-up and machine variablility. Set-up will become a much more critical operation as the technology of manufacturing moves toward flexible manufacturing systems and the use of group technology. As lots become smaller and smaller in a flexible manufacturing system, there will be more and more set-ups. Futhermore, operators will no longer have the luxury of tinkering with set-up adjustments until a minumum number of acceptable products are produced. Emphasis will be on set-ups done correctly the first time. Lill, Chu and Chung introduce a methodology which provides information on set-up error on the very first piece. They address difficulty of adjustment and handling of trend effects as well as the transition to continuous production.

Traditional statistical process control treats the mean and variance as constants and is designed to detect changes in either. Many industrial processes are characterized by repeated perturbations in the mean and the variance. Sturm, Melnyk, Feltz and Wolter propose a methodology for tracking dynamic industrial processes. Their sufficient statistic process control is actually a Kalman Filter which updates both the mean and the variance on each observation. An interesting feature of their procedures is that of detecting shifts in the mean by comparison of the short with the long term mean using box and whisker plots. They also discuss applications of the procedure to instrumentation problems, conforming to engineering specifications and process control issues. Illustrations are given using a circuit board assembly process.

Jablonsky, Davis and Hitchings illustrate how statistical process control might be used to monitor the dynamic response of a continuous process in order to discover out of control conditons. Using two crude units of an oil refinery, the combined Shewhart-CUSUM procedure is shown to be an effective tool to determine out of control conditions reliably. Continuous processes are characterized by both the normal variation in the process and the dynamic interactions between the process control loops. The authors report sucess in this environment and offer several suggestions for additional work with continuous processes.

In a computer integrated manufacturing environment, it is quite common to find serial correlation in the observations. In fact, autocorrelated data will likely be an important aspect of any statistical process control application involving sensors and on-line measurement techniques when observations are taken relatively close together in time. The paper by Yourstone deals with this problem, and shows how a discrete time series model can be used to post-whiten the observations so that conventional control charting methods can be applied to the residuals. The result is an algorithm for process control that should have wide appliability in both discrete parts and in certain kinds of process operations. The algorithm is tested using ninety simulation runs in six designed experiments.

Flaig offers adaptive control charts in which the adaptation principle links the sample size to whatever trends might develop in the data. The effect is one of substantially increasing the sensitivity of the chart with only a moderate increase in inspection. The adaptation principle is based on changing the sample size of the next subgroup based on the zone in which the sample mean of the previous subgroup lies. Flaig develops Average Run Lengths (ARL's) for the scheme and offers an economically-based model illustrating the trade-off between ARL and the Average Sample Number (ASN).

A well-known contributor to the field of multivariate quality control himself, Jackson offers an interesting review of multivariate procedures applied to quality control since

Hotelling's classic paper. He presents an overview of most of the major multivariate methods in use today including principle components, multivariate CUSUM, Andrews plots and multivariate acceptance sampling and control charts. He concludes that the field is ripe for many new developments as several areas of research have hardly been tapped.

SECTION II

Keats reminds quality practitioners that setting parts per million goals is not a meaningful activity in the absence of a means to monitor processes to determine whether or not such goals are being met. He suggests the use of the time-between-events CUSUM as an effective way of monitoring low defect processes and determining changes in either direction from the tolerable defect level. He points out that this type of CUSUM is based on the ratio of detectable to tolerable defect levels and not order of magnitude of the levels. A success rate of over 90% in distinguishing 2 from 1 defect per million is reported. Keats also advises reliability engineers engaged in the design process to be cautious of goals involving "mean time between failures" or "mean time to complete a maintenance action". As an alternative, he offers procedures based on the tails of the appropriate distributions.

The paper by Montgomery presents several examples of analyzing both location and dispersion effects from designed experiments and compares the results of this methodology with Taguchi's parameter design and signal-to-noise ratio data analysis technique. It is demonstrated that the Taguchi Methods are generally ineffective and inefficient when compared to more scientific alternatives. That is, the recommended methodology yields better discrimination and generally requires less experimentation than does Taguchi's methods. The methodology is applied to industrial data in three illustrative situations.

Mikel Harry's first case study in statistical experiment design looks at solder flux types and their relationship to solderability of components to a printed wiring board. A rather innovative aspect of this study is the use of a five point Likert scale to classify corrosion. This transformation is an attempt to overcome one of the difficulties associated with a great deal of the data associated with the electronic industries--it is often in binary form (good/bad; go/no-go). The Likert scale forces a choice in categores (five in this case) ranging from terrible to excellent and yields a dependent variable which while not normal can be applied in the analysis of variance.

Harry's second study is a designed experiment examining processes associated with solder joint cracking. Once again, a Likert scale is employed. Use of a fractional and full-factorial design is illustrated.

SECTION III

Baxley offers a compromise between time series control methodology which requires an adjustment at every sample interval and Shewhart control schemes which adjust only when an out-of-control situation exists. He looks at feedback control strategies aimed at minimizing the standard deviation of the control error. The Exponentially-Weighted Moving Average (EWMA) and CUSUM procedures are compared for use with a drifting process (simulated with an integrated moving average model). Average adjustment intervals range from 5 to 20 sample periods. The effect of deadtime is also considered.

A bivariate process capability vector is suggested by Hubele, Shahriari and Cheng. The vector has three components--observed process variation relative to that allowed by specifications (as in the traditional C_p ratio), relative location of the process and specification centers, and relative location of the maximum and minimum values in the probability contours and the specification limits. The vector is illustrated with actual manufacturing data. The vector is intended for use in any situation where two variables are to be controlled together.

Svoboda summarizes the literature associated with the economic design of control charts in the interval (1979-1989). Her review includes commentary on over 40 papers and she presents a bibliography of those papers not reviewed. She points out that although cost-based models are typically more difficult to apply relative to risk-based models, research in the last decade is primarily aimed at simplification based on approximations and the use of computer programs.

Faltin and Tucker present an intriguing view of how on-line process quality control methods will operate on the factory floor of the 1990's. In this environment, automated monitoring and response to process problems will be of major importance. They argue that simplistic automation of current methodology will not suffice in the next decade. They also discuss the expanded role of statistical process control in the new production environment and they propoe a current-technology approach for use in first generation automated plants. They conclude by delineating the form of a next-generation system and propose areas where research is needed to support the development of such a system.

SECTION I

PROCESS CONTROL APPLICATIONS

2
SPC: USE IN LOW VOLUME MANUFACTURING ENVIRONMENT

George F. Koons*
Manager - Quantitative Problem Solving

Jeffery J. Luner
Technical Specialist - Quantitative Problem Solving
McDonnell Aircraft Company
St. Louis, MO 63166

Abstract

One concern about the use of SPC in McDonnell Aircraft centered around the fact that the company is a low volume manufacturer. Concerns about the myriad number of parts, the uncountable number of part characteristics, and the short production runs of any particular part immediately raised questions about the practicality of applying SPC techniques. Using a pilot site in the Machine Shop these concerns were evaluated.

The solution was to study the process, not the product. Rather than characterizing specific dimensions on specific parts, the processes that produced those dimensions were identified. The throughput of several processes were monitored during a two-month capability study. Control charts were used to identify special causes of variation, and regression analysis was used in conjunction with a database to examine potential common causes of variation. Once the process was improved, control charts were implemented to monitor both within lot variation and longer term process variation.

*Current affiliation: Motorola Lighting, Inc., Buffalo Grove, Illinois

Background

During 1986, the McDonnell Aircraft Company (MCAIR) adopted "a 90% improvement in quality and a 40% reduction in the cost of its product by 1991" as its Significant Business Issue. These ambitious goals were selected because they will only be achieved when the company changes the way it conducts its business. One necessary cultural change is a increased use of statistical methodology in all phases of the business. This paper discusses an application of Statistical Process Control (SPC) in the MCAIR Machine Shop.

One concern about the use of SPC in MCAIR was that the company's status as a low volume manufacturer. Concerns about the myriad number of parts, the uncountable number of part characteristics, and the short production runs of any particular part immediately raised questions about the practicality of applying SPC techniques.

MCAIR delivers about 15 aircraft per month. These are extremely complicated pieces of equipment, and each is made up of several hundred thousand parts. Even a relatively simple part may have as many as 40 engineering dimensions that must be met to extremely close tolerances, typically +/-0.010 inch. Many of these parts are made in the MCAIR Machine Shop.

Within the Machine Shop, a small cell of three identical, three-axis numerically controlled vertical milling machines were selected for the pilot project. They are known as manual data input (MDI) machines. These machines manufacture about 150 different parts. Because the MDI machines are numerically controlled, the parts tend to be among the more intricate ones manufactured in the Machine Shop, and the different parts are vastly different in appearance.

In addition to the diversity of parts, the pilot situation was further complicated by short production runs, which are attributable to the low production rates. Typical lot sizes are 15 to 25 units, and a job may not reappear in the shop for several months. In the future, lot sizes will become even smaller as MCAIR's inventory reduction initiatives are implemented.

Straightforward application of Statistical QUALITY Control in this type of manufacturing environment is futile. For example, if it were necessary to monitor, even a couple of dimensional characteristics on each of the 150 parts, the number of charts used would be in the hundreds, and the administrative task would be difficult. The process' capability of achieving each dimension would have to be individually determined. Since each part is manufactured only three or four times a year, this would take years to complete.

If the SPC strategy were to create a new chart for every production lot, the run would be completed before enough data were collected to assess the state of the process. In addition, while each run could be in statistical control, differences in average dimension could exist between lots, or within lot variability could differ from run to run. These situations would be nearly impossible to to detect if separate charts were maintained for each run. At best, this approach would indicate which members of a lot differ from the other members. However, it is unlikely that this strategy would lead to process improvements that would continuously reduce dimensional variability of ALL parts.

The approach we followed was to study the process, not the product being produced. Rather than characterizing specific dimensions on specific parts, the processes that produced those dimensions were identified. The investigation, then, focused on whether the process was operating consistently and capably.

Data Collection Phase

Dimensional characteristics were monitored during a two-month data collection period. In all 85 characteristics were selected. For some of the parts more than one characteristic was monitored. These characteristics were selected based on engineering considerations which deemed them critical to the performance of the aircraft. However, rather than studying the selected dimensions individually, we elected to study the processes that created the dimensions. In other words, we applied Statistical PROCESS Control, not Statistical QUALITY Control.

The MDI machines perform six distinct operations: drilling, counter-sinking, reaming, slotting, end-milling, and side-milling. Every characteristic on each part is produced by one of these processes. When the 85 monitored characteristics were grouped according to the process that produced them, end-milling and side-milling were found to have predominated. The statistical analysis of the end-milling data will be discussed in detail.

The manufacturing lot was selected as the rational subgroup. Items within a lot are made under essentially the same manufacturing conditions. For example, if the stock is from the same batch, only one machine is used. Although multiple operators will be involved if the job runs into the next shift, usually the same machining tools are used throughout the run, etc. Choice of the lot as the rational subgroup, therefore, minimizes the number of factors affecting the dimensional variability within a lot and maximizes the chances of uncovering assignable causes when comparing results from different subgroups.

Rather than sampling only a few pieces from the lot, MDI operators measured the predetermined characteristics on every piece and recorded the data. This strategy was used because time was not critical and because trends, due to tool wear, were thought to be a possibility. This decision created subgroups of unequal sizes and made the control charting and the statistical analysis somewhat more complicated than if the subgroups had been the same size.

As each lot was completed, the data were summarized by calculating the average and variance of the individual readings. Prior to the calculations, the data were "adjusted" by subtracting the nominal dimension from the actual dimensions. For example, if the nominal dimension were 0.352 inch and the measured dimension were 0.349 inch, the "adjusted" measurement would be -0.003 inch. The transformed data allowed direct comparisons between the different subgroups since each characteristic had its own nominal dimension. The variance, not the range, was used as a measure of dispersion because the latter is an inefficient statistic for subgroups as large as those in this study.

Anticipating a subsequent analysis for "common cause" effects, several other factors were noted for each subgroup. They were:

machine number:	72,73,74
material:	Aluminum, Titanium
position:	located to reference
	located to previous cut
stock:	Extrusion, Milled material
process:	end mill, side mill, etc.
subgroup size:	ranged from 10 to 58

These, along with the average and variance of the dimensional deviations from nominal and descriptive information, such as, work order number and date, were posted to a data file.

Special Causes Of Variation

After the data collection phase, a control chart analysis of the results was conducted. As always, the first step was to evaluate the process output's consistency.

Subgroup variances were displayed on a S^2 chart, Figure 1. Sample variances were used because they provide unbiased estimates of the population variance. The center line, $\overline{S}^2$, is the weighted average of the subgroup variances. The width of the three-sigma control limits varies because of the differences in subgroup sizes, n_j. The control limits become narrower as the subgroup size increases. The limits for the variance chart were calculated using the chi-squared distribution.

$$\text{Lower control limit} = \chi^2_{0.999}\frac{\overline{S}^2}{(n_j - 1)}$$

$$\text{Upper control limit} = \chi^2_{0.001}\frac{\overline{S}^2}{(n_j - 1)}$$

Four subgroup variances exceeded the upper control limit. The production sequence plots of three of these indicated the first pieces made were well below nominal dimension and that

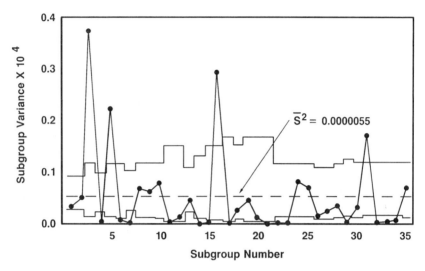

Figure 1. Initial Variance Control Chart
            MDI End-Milling Process

process adjustments were made by the operator, Figure 2. Since an assignable special cause, set-up, was identified, the affected data points were deleted from the analysis. The fourth observation that was out of statistical control, subgroup #16, was caused by an adjustment made by the operator at the shift change. This event had been noted on the data collection chart. The subgroup was excluded from the analysis.

Several of the subgroup variances were BELOW the lower control limit, an indication of significantly lower variability in comparison to the other subgroups. It is equally, perhaps even more important, to isolate assignable causes when this phenomenon occurs. However, recognizing that deletion of the observations affected by set-up difficulties would result in an overall downward shift of the control limits, these subgroups were not critically examined during the initial analysis.

The control chart analysis was repeated two additional times. After the second iteration, a few more individual data points were deleted from the study because of apparent set-up discrepancies. The final variance control chart appears in Figure 3. A few points still exceeded the upper control limit,

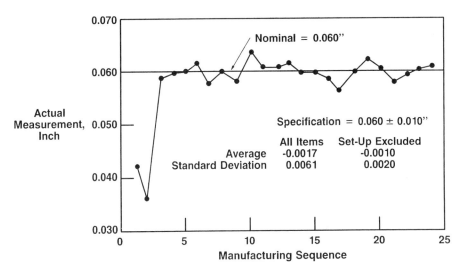

Figure 2. Example of Corrected Set-Up Situation
MDI End-Milling (Subgroup No. 3)

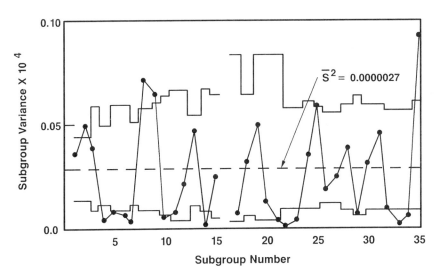

Figure 3. Final Variance Control Chart
MDI End-Milling Process

and several points were below the lower control limit of the variance chart. One plusable explanation for the "low" points on the variance chart might be measuring error. Conceivably operators, who at this point were still learning about SPC, could have been rounding-off readings. Measurement variation would have been understated if that had been the case. However since no additional assignable, special causes could be found to explain the apparent discrepant observations, the control chart analysis was completed.

The average subgroup variance was used to calculate three-sigma control limits about the x-bar chart center line, the weighted average of the subgroup averages. The width of these limits were also related to subgroup size.

$$\text{Lower control limit} = \overline{\overline{X}} - 3 \sqrt{\frac{\overline{S}^2}{n_j}} \tag{1}$$

$$\text{Upper control limit} = \overline{\overline{X}} + 3 \sqrt{\frac{\overline{S}^2}{n_j}} \tag{2}$$

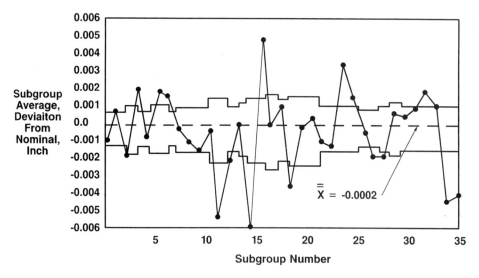

Figure 4. Initial X-Bar Control Chart
            MDI End-Milling Process

The subgroup averages were also out of statistical control (Figure 4). Sequential time plots of individual pieces in the out-of-control lots typically displayed consistent dimensions, but, in a STATISTICAL sense, at an unusually large distance from the nominal (Figure 5). All pieces in these lots were well within engineering tolerances. However, since the first pieces were within specification, no machine adjustment was deemed necessary. This phenomenon sparked discussions about whether "within tolerance" or "uniformity about target" should be the basis for quality evaluation.

Plots were then made to investigate the possible effect of tool wear. The average dimensional deviations from nominal were calculated by the sequence in the manufacturing run (Figure 6). That is, the average deviation was calculated for all start-up pieces, all second pieces, etc. Any start-up pieces that were discarded during the variance chart analysis were eliminated from these calculations. The analysis was not continued beyond the 24th item, as very few lots consisted of more than that many units. A drift in dimensions is apparent.

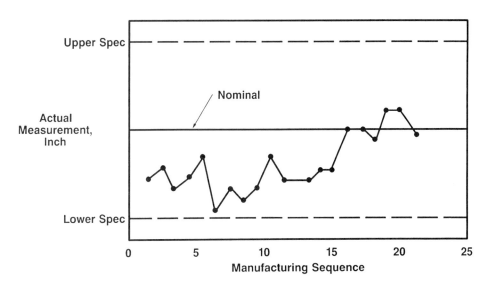

Figure 5. Example of Within Specification But Not in Statistical
Control MDI End-Milling (Subgroup No. 35)

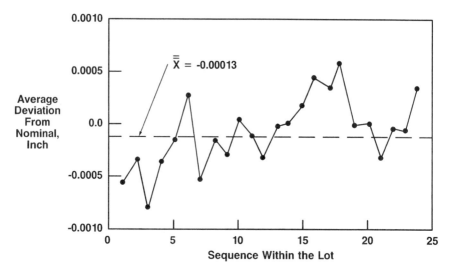

Figure 6. Effect of Tool Wear MDI End-Milling Process

By this time, three major improvement activities had been identified. Two were technical; improving set-up procedures and monitoring dimensional changes to determine when process adjustments should be made to compensate for tool wear. The third, and most difficult, was managerial; adopting the philosophy of uniformity about target value and being willing to make minor process adjustments when the product is out of statistical control, even though it is within engineering specification.

Common Cause Of Variation

The next step was to see if any operating factors had a systematic effect on the subgroup variability. Several potentially influential variables had been noted as the data were collected. These were machine number, type of material, type of stock, and location of the studied dimension relative to a known reference location on the fixture. Multiple regression analysis was used to examine the data. The four operating factors were the independent variables, and the subgroup standard deviation was the response variable. Because all the independent factors were categorical, dummy variables were used in the regression model. Two dummy variables (X1 and X2) were required to study the differences between the three machines, and one (X3) was needed to assess any difference between located positions.

Preliminary analysis of the data revealed parts had not been machined from all four possible combinations of material type and type of stock during the study period. Aluminum extrusions and milled stock were used, but all the titanium parts had been made from milled starting stock. Including a dummy variable to assess the effect of material and a second to investigate differences between types of stock would have introduced intercorrelated independent variables. Instead, three material/stock combinations were defined, and two dummy variables (X4 and X5) were included in the regression model to investigate any differences between them.

Since no interactions between the independent factors were anticipated, an additive regression model was assumed. The logarithm of the subgroup standard deviation was used as the response variable. When working with the standard deviation, this transformation is typically used to satisfy the normality assumptions of the statistical procedure.

$$\log SD = b_0 + b_1 X_1 + b_2 X_2 + b_3 X_3 + b_4 X_4 + b_5 X_5 + e \quad (3)$$

where,

(X_1, X_2) = (0,0) if machine 72 had been used
= (1,0) if machine 73 had been used
= (0,1) if machine 74 had been used

X_3 = 0 for located dimensions
= 1 for resultant dimensions

(X_4, X_5) = (0,0) if milled, aluminum stock had been used
= (1,0) if aluminum extrusions had been used
= (0,1) if milled, titanium stock had been used

e = error term that satisfies the usual assumptions of independence and normality

TABLE 1.
ANALYSIS OF VARIANCE
COMMON CAUSE REGRESSION MODEL

SOURCE	DF	SS	MS	F
Model	5	0.5273	0.1055	1.38
Error	28	2.1429	0.0765	
Corr Total	33	2.6702		

The lack of statistical significance of the regression model indicated that within-subgroup variation was not related to the operating factors in the model, Table 1.

The actual differences between the various categories of the operating factors are shown in Table 2.

TABLE 2.
EFFECT OF OPERATING FACTORS ON DIMENSIONAL
VARIABILITY

FACTOR	AVERAGE STD DEV	
	LOG 10	INCH
Machine: 72	-2.95	0.0014
73	-2.91	0.0014
74	-2.93	0.0015
Surface: Located	-2.91	0.0015
Resultant	-2.98	0.0012
Material: Milled Aluminum	-3.01	0.0012
Extruded Aluminum	-2.94	0.0014
Milled Titanium	-2.74	0.0020

Corrective   Action

The General Foreman of the MDI cell developed a set of standard set-up procedures to be used by all operators. This greatly reduced the occurrence of large discrepancies from nominal for the first piece in a lot.

In addition, statistical control charts are being used to determine when adjustments are needed to compensate for small set-up departures from nominal and to monitor dimensional drift because of tool wear. Operators are maintaining an X-chart (a control chart for individuals) and a moving range chart as each piece is produced. Probabilistic rules are being used to indicate when the process is out of statistical control, and a machine adjustment is needed to return the process output back to the target value.

To ensure long-term process consistency, the individual measurements are statistically summarized as each lot is completed. After culling out any special cause effects, the subgroup average and variance are posted to control charts. Any point, subgroup, falling outside the three-sigma control limits is investigated to determine the reason for the lot being atypical.

The logic for this monitoring approach is shown in Figures 7 (a) and (b) which represent the flow charts of the procedure.

Process   Capability

Since Statistical Process Control has not yet been fully achieved, an actual process capability index cannot be reported. However, assuming that will occur, we can use the existing data to speculate about the index. The average subgroup variance was 0.0000027 sq. inch. The estimated standard deviation, therefore, is 0.0016 inch. For a dimension having the most common tolerance encountered on parts made in the MDI cell (±0.010 inch), the capability index would be:

$$C_p = \frac{\text{Tolerance}}{6 \text{ Standard Deviations}} = \frac{0.020}{6 * 0.0016} = 2.08 \tag{4}$$

Tighter tolerance greater the effect of the machine greater cost involved TIGHTENING of deviation

AVG 1.03

out of limits

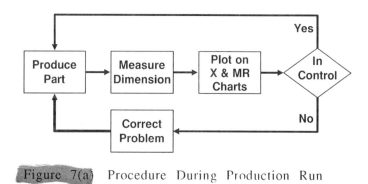

Figure 7(a)   Procedure During Production Run

Figure 7(b). Procedure After Run Concluded

⟶   For the tightest tolerance cited for end-milled dimensions, ±0.005 inch, the capability index would be 1.04.

An important point to note is that the index would be applicable to all dimensions produced on the MDI machines using the end-milling process. This is true since, to date, no differences in process capability have been found that are related to material type, relative location of the cut, machine number, or any other operating factor. Continuing analysis of the data may eventually determine that the process capability differs depending on some external, uncontrollable factor. If that were the case, perhaps two or three capability indices would be required to describe the process. In any event, having only those few measures of capability would still greatly simplify an SPC activity when compared to the approach of monitoring each dimensional characteristic as if it were unique.

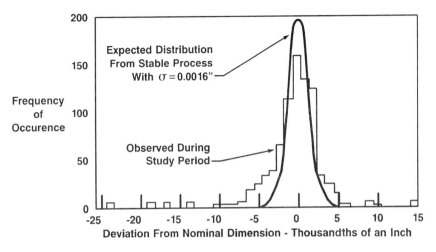

Figure 8. Comparison of Actual Product With Product
           From a Stable Process. MDI End-Milling Process

Figure 8 shows the potential gain in the consistency of dimensional characteristics once the process is brought under statistical control. The histogram shows the distribution of 815 dimensional deviations from nominal values that were observed for end-milled characteristics during the data collection phase. A normal distribution of dimensional discrepancies that is centered on target with a standard deviation of 0.0016 inch is superimposed on the distribution of observed data.

Conclusion

MCAIR's pilot Statistical Process Control activity in the MDI area has demonstrated that the traditional tools are readily adaptable to a low-volume manufacturing environment. The keys to using the tools are to concentrate on the process, not the product, and to work with deviations from nominal dimensions rather than the actual reported dimensions.

Operating factors can be statistically evaluated to determine which, if any, have a significant effect on the process capability. Modifications to the process may be possible to eliminate these effects. Influential process factors that cannot be changed, such as, types of raw material will have to be treated as a family. However, Statistical Process Control can still be implemented on a much broader basis, and therefore much more simply, than if each part were considered to be unique.

3
STATISTICAL SETUP ADJUSTMENT FOR LOW VOLUME MANUFACTURING

M. H. Lill, Contract Project Manager
Yen Chu, Member of Technical Staff
Ken Chung, Member of Technical Staff
Central Engineering Laboratories, FMC Corporation
Santa Clara, CA 95052

Abstract

Statistical methods are proposed to estimate, starting with the first piece produced, the most probably correct adjustment to make in a machine setup to produce pieces as close to the desired dimension as possible. Methods are also presented to minimize the number of adjustments recommended and to anticipate the effects of a known trend such as abrasive tool wear. Charts are not required but may be used to illustrate the results of recommended adjustments.

Need for a new approach to statistical process control

Development efforts in Flexible Manufacturing Systems for low volume production are introducing machining centers and manufacturing cells in which toolpaths, tool settings and/or fixture settings are controlled by a computer. A natural extension of this development is to have automatic inspection with data fed back in real time to the control program so the computer can correct for initial setup errors or drift. When that capability is realized, it will be necessary to include a process which can respond immediately, after the first piece if necessary, and still avoid overcontrol. Some systems are attempting to address this need with conventional Statistical Process Control, but this does not promise much success in low volume production.

Conventional SPC does not work well in low volume manufacturing

The technology of Statistical Process Control (SPC) is driven by the need to reduce the number of inspections required with a large number of parts so that statistically sound decisions about a given parameter may be made by qualified observers. The primary methodology of SPC entails selecting for inspection a number of representative subgroups, determining the grand mean of the subgroups as an estimator of the mean of the population, determining the range or variance within each subgroup and calculating control limits for the means of all subgroups which represent an acceptable area under the normal distribution curve. These limits are defined as "natural" control limits and the only decision the SPC process makes is whether the parameter is thereafter "in statistical control" or "out of statistical control".

It is left up to the qualified observer to decide whether to make an adjustment to the mean of the process or to look for an assignable cause for an out of control condition. Various charting techniques have been devised to help the observer make decisions, so that charts have become the essential tool of SPC even though calculations alone could provide the same information.

In low volume manufacturing, there are often not enough pieces produced in one run to establish control limits before the run is complete. Consequently, some authors (Koons and Luner [1988]; International Quality Institute [1987]) have presented methods for combining data from more than one run, even from different parts, to characterize the variability of a given machine operation. Their objective was to improve the performance of the entire machine shop, rather than to feed back in real time, to the operation in progress, any adjustments which would improve conformity of the parts being produced. It is possible to do both of those things with Statistical Setup Adjustment.

Statistical setup adjustment is non-conventional SPC

Statistical Setup Adjustment (SSA) applies only to operations which have a measurable means of adjustment which directly affects the average outcome of the operation. It is useful only when the operation has a significant inherent variance, such that the true central tendency is not apparent until several pieces have been produced. It does not rely on charts, although charts may be used to illustrate its functions.

SSA has 2 essential purposes: first, to recommend adjustments (tool setting, fixture location or tool path) as soon as possible for each dimension-producing operation so the chance of producing a part out-of-limits is as low as possible; second, to provide statistical information about the operation which can be recorded and utilized in planning similar operations on the same machine or in decisions about maintenance or process changes. Higher priority must be given to the first purpose because the nature of small batch production imposes a double penalty for reject parts: cost of salvage or replacement parts, and schedule slippage when the expected number of good parts is not available.

Concept of potential populations

SSA is based on the concept that each repetitive machine operation is a unit in a potentially infinite population with some variance about a discrete central tendency. Each dimension on a part has some error representing the algebraic sum of the operation variation and the setup error. Likewise setup is a repetitive operation, although in each small production lot there should be only one initial setup condition. The error in each setup is apparent only as the departure, from the desired dimension, of the average of the parts produced from that setup. However, if a large number of small batches of the same part eventually are produced, the aggregate of setup errors would represent a population with a central tendency close to zero and a variance related to the difficulty of making a precise setup. This concept, coupled with an assumption that the potential populations will have near-normal distribution characteristics, allows us to use statistical methods to estimate the portion of dimensional error due to setup error, based on given variances for the potential populations.

Estimation of setup error after the first piece

It is possible to estimate the variance of the machine operation from past performance. If machine capability has been evaluated, or if SSA has been used on a previous run of the same operation, variance is known. It is also possible to estimate the variance or standard deviation of the setup error, based on expert opinion. The shop supervisor, or the setup man, can rate the setup as Very Hard, Normally Difficult or Very Easy. These perceptions can be equated to dispersions representing some fraction of the available print tolerance.

From available information and experience, it should be possible to calculate an adjustment after the first production piece is measured which has the greatest probability of being correct; and/or an acceptable small probability of causing a greater error in the centrality of the setup adjustment than already exists.

Simplest combination of distributions of error

If the machine operation and the potential setup errors have distributions with equal variance, it can be seen intuitively that the error (X_1) measured on the first piece is most probably due 50% to setup error and 50% to machine variability. This is readily confirmed by comparing the possible combinations of cells in histograms representing each population (see Figure 1).

In Figure 1, the equal normal distributions of 10,000 machine operations and 10,000 setups are plotted and divided into histograms of equal increments of error. First piece error is the sum of machine error and setup error, and is presented on a nomograph line between the 2 distributions. The relative probability of any assumed value of setup error being correct is the product of the content of its cell and the content of the machine error cell which corresponds to it in the nomograph, divided by the product of all possible combinations of both entire populations. In the illustration, the product 425 x 425 = 180,625 exceeds the product of any other available combination, confirming in this case that 50% of the piece error is most probably due to setup error.

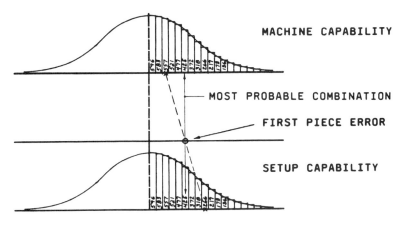

Figure 1. Combination of Equal Variations

As shown, the total error on the first piece is .0024". If an adjustment in setup of .0012" is made, there is .88 probability it will be in the right direction, .77 probability it will not make matters worse, and .08 probability it will be correct within ±0001". If an adjustment equal to the total error, .0024", were made, there is only .48 probability it will not make matters worse, and.05 probability it will be correct within .0001".

Combination of unequal normal distributions

If the variance of potential setup errors is much larger than the machine operation variance, the most probable portion of 1st piece error due to setup again can be determined by the histogram method (see Figure 2).

In Figure 2, the unequal normal distributions of 10,000 machine operations and 10,000 setups are plotted. In this illustration, the product 663 x 318 = 210,834 exceeds the product of any other available combination, indicating that in this case approximately 2/3 of the piece error is most probably due to setup error. As shown, the total error on the first piece is again .0024". If an adjustment in setup of .0016" is made, there is .98 probability it will be in the right direction, .93 probability it will not make matters worse, and .11 probability it will be correct within ±0001". If an adjustment equal to only 1/2 the

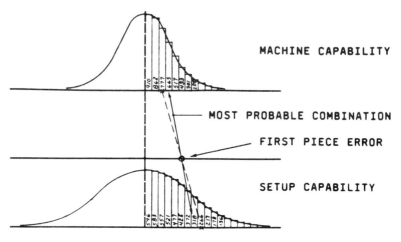

Figure 2. Combination of Unequal Variations

total error, .0012", were made, there is .95 probability it will not make matters worse, but only .09 probability it will be correct within .0001".

Direct calculation of best adjustment value after first piece

Likewise the most probable setup error can be calculated after the 1st piece if values are given for the standard deviation of machine variability (sigma$_m$) and the standard deviation of all potential setup errors (sigma$_s$). The normal probability distributions of machine and setup variations are (see Hoel, Port and Stone [1971]):

$$Y_m = (2 \ pi)^{-.5}(sigma_m)^{-1} \ e^{-(X_m - u)^2/2(sigma_m)^2}$$

$$Y_s = (2 \ pi)^{-.5}(sigma_s)^{-1} \ e^{-(X_s - u)^2 /2(sigma_s)^2}$$

where: X_m = any specific value of operation error
        X_s = any specific value of setup error
        u   = central value

Assuming the central value of both distributions, $u = 0$:

$$\text{Total error } E = X_m + X_s \;;\; X_m = E - X_s$$

then let: $c = \text{sigma}_s/\text{sigma}_m$

The combined probability distribution of $F(X_s)$ is:

$$Y_m * Y_s = (2 \text{ pi sigma}_m \text{sigma}_s)^{-1} e^{-Z[1/2(\text{sigma}_s)^2]}$$

where: $Z = c^2(E - X_s)^2 + X_s^2$

Maximum $F(X_s)$ occurs at minimum Z; then, differentiating Z:

$$(dZ/dX_s)/2 = -c^2(E - X_s) + X_s$$

and, setting the differential equal to zero:

$$X_s = Ec^2/(c^2 + 1)$$

The best adjustment (A_1) is the difference of the desired dimension (D) and the measured result X_1, multiplied by the ratio X_s/E:

$$A_1 = (D-X_1)c^2/(c^2+1) \tag{1}$$

where: $c = \text{sigma}_s/\text{sigma}_m$

Sigma$_m$ can be estimated by the sample standard deviation s_m of previous machine runs. Sigma$_s$ is really not possible to determine, of course, although it must be related in a practical way to the print tolerance. It must also be related in a practical way to the apparent difficulty of making a precise setup or tool setting before a piece measurement is available. A factor (U) may be assigned to the apparent difficulty. A better expression for A_1 may be:

$$A_1 = (D-X_1)(T/Us_m)^2/((T/Us_m)^2+1) \tag{2}$$

where: T   = total print tolerance on dimension D
        U   = an empirical constant, related to the
              difficulty of precision setup:

            Very Hard: U = 5
            Normally Difficult: U = 8
            Very Easy: U = 12

     Intermediate values may be chosen.   The value T/U is
actually an estimate of speculation at the value of sigma$_S$.   In the
case of a very difficult setup, one would expect the setup error
to exceed 20% of the total tolerance or 40% of the bilateral
tolerance about 1/3 of the time.

Calculating adjustment after second and later pieces

     When the 2nd and later pieces are run and measured, a new
form of information becomes available: X-bar (mean result).   X-
bar is an unbiased estimator of the true setup value which
becomes statistically more precise as the number of pieces run
(the sample size, n) increases.   If the setup and first pieces were
repeated a very large number of times (t), the errors between X-
bar and the true setup value would form successively narrower
normal distributions over t trials for each value of (n) with
variances proportional to 1/n and standard deviations
proportional to 1/(sq.rt.of n), as illustrated in Figure 3.   Then,
modifying Equation (2), after n pieces the best adjustment value
(A_n) may be estimated by the expression:

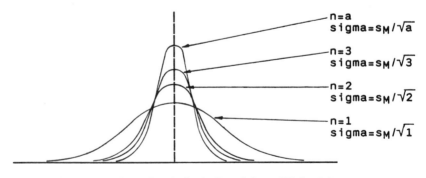

Figure 3. Improving Statistical Precision With (n)

$$A_n = (D - \text{X-bar}_n)(T/Us_m)^2 n/(((T/Us_m)^2 n)+1) \qquad (3)$$

As n becomes large, A_n approaches (D - X-bar). While n is small, X-bar can be quite volatile; but overcorrection is minimized by the calculated adjustment being some fraction of (D - X-bar).

NOTE: In order to maintain a continuous relationship between X- bar and the previously taken data, in spite of the step inputs represented by any setup adjustments, X-bar$_n$ should be corrected after any adjustment and each succeeding value of X-bar should be calculated as though all adjustments had been made before the 1st piece was run, as follows:

$$\text{X-bar}_n' = \text{X-bar}_n + A_n \qquad (4)$$

$$\text{X-bar}_{n+1} = ((\text{X-bar}_n' * n) + X_{n+1})/(n+1) \qquad (5)$$

Moderation in applying adjustments

As long as data is taken from each piece, an adjustment may be calculated, no matter how small. To avoid frequent insignificant adjustments, a threshold must be set below which the calculated adjustment (A_n) is ignored. Such a threshold may be based on the total tolerance versus the apparent need for very precise centering of the setup. Since the maximum acceptable s_m is usually considered equal to T/6 (random machine error alone would exceed the tolerance in 3 pieces out of 1000), the minimum desirable adjustment (m) may be defined as the absolute difference between $6s_m$ and the total tolerance (T), multiplied by a suitable constant (M), as follows:

$$m \quad \pm M|T-6s_m| \text{ (rounded to L dec. plc.)} \qquad (6)$$

where: M = empirical constant based on
                 relative ease of adjustment
          T = total print tolerance.
          L = selected no. of dec. plcs. (usually 4)

When $A_n < m$, the recommended adjustment (A_r) should be arbitrarily set to 0 to avoid confusing calculations. Suggested values for M :

Difficulty of Adjustment	Values of M
Automatic	.1
Easy	.2
Normal	.3
Hard	.4

Intermediate values may be chosen. M represents the fraction of the margin between $6s_m$ and the level of tolerance which is acceptable in order to avoid too frequent adjustments. Machine selections where s_m is more than T/6 are are disregarded.

In addition to the threshold set to avoid frequent small adjustments, there is a practical limit: the smallest adjustment increment (A_m) that can be made. This value depends on the means of adjusting the setup dimension, and must be selected by someone familiar with the particular machine and tooling. Provision should be made in any calculation program for inputting this value and for rounding all recommended adjustments to a multiple of it.

Avoiding early false adjustments

Early recommendations for adjustment are suspect of being triggered by chance occurrences of machine errors which are near the maximum in the population, when additive setup error is really not present. For the first piece, a minimum total error threshold may be set at $3s_m$, which would allow such a false recommendation less than 3 times in 1000 lots. For the 2nd and later pieces, the probability that the calculated adjustment will approximate the actual setup error improves rapidly, so it is appropriate to decay the minimum error threshold until it is no longer significant. A good way to do this is to diminish it by a factor of 1/(sq.rt. of n). This threshold permits recommendation of early adjustments only after rather bad pieces, which generally agrees with the perception of the operator. It significantly reduces the incidence of false adjustments without a serious increase in parts out-of-limits due to delayed adjustments.

Transition from startup to continuous production

At some number N of pieces run, X-bar will have a sufficiently high probability of estimating the setup error within the adjustment threshold (m) already calculated that X-bar itself may be used for the recommended adjustment. At that time, Equation (3) may be replaced by an expression based solely on X-bar. N may be calculated so that the probability of X-bar differing from the true mean of the potential population by more than the threshold adjustment (m) is less than .003 after N pieces are run. A confidence level of 0.997 requires an area under the normal distribution curve bounded by ± 3 sigma, giving an expression for N based on (m) and (s_m):

$$m = 3s_m/(\text{sq.rt.of } N) \quad \text{or:} \quad N = (3s_m/m)^2 \qquad (7)$$

Adequate response to trend in piece errors

If there is any trend or drift in the average value of the produced dimension (as from tool wear), X-bar based on equal weight for all pieces already run will be inhibited from following the trend. To permit X-bar to follow any long term trend, and yet be stabilized adequately against short term variation, an exponentially weighted moving average (EWMA) based on a selected base number (K) may be substituted for X-bar after K pieces are run:

$$\text{X-bar}_n'' = ((\text{X-bar}_{n-1}') * (K - 1) + X_n)/K \qquad (8)$$

and:

$$A_n = D - \text{X-bar}_n'' \qquad (9)$$

where:

$\text{X-bar}_n''$ = weighted X-bar_n to be used for calculating A_n
$\text{X-bar}_{n-1}'$ = adjusted X-bar_n after previous piece.
K   = a factor based on s_m and the threshold (m), explained below:

For the case where there is little or no trend and the process mean is already well centered on the desired dimension, K may be calculated so that X-bar$_n$" will vary only enough to exceed m, and thus cause an unnecessary adjustment, about 3 times in 1000 pieces:

$$K = (3s_m/m)^2 \qquad (10)$$

It is apparent that the value for N described in the preceding paragraph is the same as the value for K, and n=K may be used for both transitions. However, when the total tolerance (T) is not much greater than $6s_m$, m becomes quite small and so K becomes large in relation to the typical number of pieces in a small batch. Consequently, it appears practical to set a maximum for K at 11. This permits X-bar$_n$" to follow any steady trend without too much lag and still protects quite well against unnecessary adjustments caused by random sequences of variations. For convenience in measuring and compensating an anticipated trend in dimensions, as explained below, it is better to use a fixed value of K = 11 in all cases and a value for the trend factor based on the accumulating trend per 10 pieces.

Anticipating effects of a known trend

Some operations may have a well-known, documented trend in error values due to some unavoidable cause, such as abrasive tool wear. When this is true, the best method of compensating without making a large number of extremely adjustments is to add a bias to the adjustment recommendations of the program so that they lead (D - X-bar$_n$") by an amount (B), which is defined as the accumulating amount of the known trend in 10 pieces. Equation #2 for the 1st piece is then modified as follows:

$$A_1 = (D - X_1)(T/Us_m)^2/((T/Us_m)^2 + 1) - B \qquad (11)$$

As additional pieces are run, the trend, if any, is accumulating. By the time K=11 pieces are run, if no adjustment has been made, a total bias equal to 2B is necessary to compensate for already accumulated drift and then lead the trend by an equal amount. Equation #9 is then modified as follows:

$$A_n = D - \text{X-bar}_n'' - 2B \qquad (12)$$

For the period between the 1st piece and the 11th piece, a proportionally increasing bias should be added to the recommended adjustment. Equation (3) is then modified as follows:

$$A_n = (D - \text{X-bar}_n)(T/Us_m)^2 n/(((T/Us_m)^2 n) + 1)$$
$$-(n + n/10)B \qquad (13)$$

The lag effect of the EWMA (Eq. (8)) at $K=11$ permits a following error in X-bar$_n''$ equal to (B); so, with the leading biased adjustments, the net effect of the trend on the average produced dimensions tends to be neutralized. After any adjustment, the normal variations in X-bar$_n''$ continue to combine with the accumulating trend error and the adjustment bias until the threshold is exceeded and another recommendation for adjustment is generated. Each adjustment has a particularly high probability of being in the right direction to compensate the trend and a low probability of combining with normal variation to cause a part out-of-limits (bias applied too early) or of being too late to prevent the trend from causing a part out-of-limits.

Avoiding frequent adjustments caused by known trend

When a significant trend is present the accumulating offset may equal the threshold adjustment value every few pieces. The only fair way to minimize the number of adjustments is to increase the adjustment threshold value. To do so automatically, the absolute value of the trend in 10 pieces (B) may be factored into the calculation for the threshold (m), modifying Equation #6 as follows:

$$m = M(|T - 6s_m| + |(K-1)B/2|) \qquad (14)$$

It must be noted that s_m should be calculated from data that has the trend removed. A computer program operating on tabular data can do so with great precision. Also, T (total tolerance) should be larger than $6s_m + B$ to offer a high probability of keeping parts within limits without making a large number of small adjustments.

Charts for illustrating results of adjustments

Although recommended adjustments are entirely calculated
and do not depend on chart interpretation, a simple charting
method will show both the quality of the dimensions produced
and the efficiency of the adjustment algorithms and the
constants selected (see Figure   4).

Charts such as this may be produced automatically by a
computer which also controls an automated process.   This chart
is similar to "Charts for Individuals" or "Run Charts" with the
addition of a plot of X-bar with each new piece.   The central line
is the intended dimension; or zero if error values only are
plotted.   Actual piece dimensions (or the errors represented)
are plotted as X's.   X-bar (or X-bar") values are plotted as
circles.

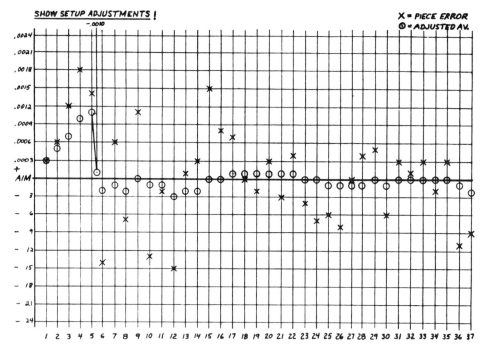

Figure 4. Typical AIM Chart

When an adjustment is made, a vertical line is drawn between the current column and the next part column, the last X-bar value is corrected by the amount of the adjustment, and the corrected X-bar value is plotted on the vertical line. A diagonal line is drawn from the X-bar value before adjustment to the corrected X-bar value after adjustment. The amount of the adjustment is noted at the top of the vertical line.

Conclusions

The method presented here is the best way found by the authors to bring a machining process with significant variance into best adjustment in the least possible number of pieces. It requires 100% inspection in real time of the first several pieces, which, if done manually, may be costly and cannot be recommended for every operation. In any case where there is a high risk of violating tolerance because of machine variability and/or drift, there is an obvious advantage to measuring in real time versus completing a batch and 100% inspecting to sort out scrap. After a process is under control, various sampling plans may be used to reduce the amount of inspection. The same calculations may be used to continue monitoring the process if regular sampling is employed.

The program to implement this method is simple enough to utilize a pocket calculator, such as an HP41, to provide the ultimate in portability and the least intrusion on the factory floor. However, it is rigorous enough to justify its use in dynamic machine control with automatic inspection feedback. It can easily be adapted to existing SPC systems where electronic reading inspection tools deliver data to computer files for processing.

The method was conceived with the purpose of eliminating dependence on charts. However, charts may have real value for illustrating causes of quality problems and the effects of corrective measures. Values calculated by the program can be used quite effectively to make charts for illustrative purposes (see Figure 4, referred to above). Where computer files are generated, charts may be delivered on demand for documentation or analysis.

References

Hoel, Port and Stone, (1971) Introduction to Probability Theory, Houghton-Miflin.

International Quality Institute, Inc., (1987) SPC For Short Production Runs: Seminar, Detroit.

Koons, G. F. and Luner, J. J., (1988) SPC: Use In Low-Volume Manufacturing Environment:, ASQC Quality Congress, Dallas.

4
SUFFICIENT STATISTICAL PROCESS CONTROL: MEASURING QUALITY IN REAL TIME

George W. Sturm, Associate Professor
Department of Mathematics
Grand Valley State University
Allendale, MI   49401

Steven A. Melnyk, Associate Professor
Department of Management
Graduate School of Business Administration
Michigan State University
East Lansing, MI   48824

Mona A. Yousry
60TR3 61400
AT&T Technologies
P.O. Box 900
Princeton, NJ   08540

Carol J. Feltz
60TR3 61400
AT&T Technologies
P.O. Box 900
Princeton, NJ   08540

James F. Wolter, Associate Professor
Department of Marketing
Seidman School of Business
Grand Valley State University
Allendale, MI   49401

Abstract

For any process control system to be successful, there must ultimately be a fit between the manufacturing environment, its operating characteristics and the process control system. Many current process control techniques such as Statistical Process Control (SPC) are primarily for manufacturing environments which range from the job shop to batch manufacturing environments. Procedures such as SPC are hard pressed when used in a high volume, highly automated, information intensive, short cycle manufacturing environment. Maintaining quality in such a setting requires a new procedure. This paper explains the shortcomings of SPC in information intensity manufacturing domains and introduces such a process control procedure, Sufficient Statistic Process Control (SSPC). SSPC, drawing extensively on empirical Bayesian techniques, reduces the amount of data that must be collected, analyzed and stored down to a few *sufficient* statistics. The result is a system that is well suited to the requirements of this manufacturing setting. This paper has two objectives. The first is to examine the process of fitting the process control system to the manufacturing environment. The second is to *briefly* describe the major features and summarizes the operation of Sufficient Statistic Process Control.

Introduction

Process control is that component of the overall quality management system responsible for maintaining existing quality standards by monitoring and analyzing quality characteristics while the manufacturing process is operating. It attempts to ensure that quality is present in the system, rather than inspected in afterwards. the process control system consists primarily of two elements. The first are the techniques used to collect, store, summarize, and analyze quality-related data. The second is the infrastructure by which the information generated by the techniques is used to correct quality problems (i.e., closing the loop).

Because process control system is so closely linked to the operation of the manufacturing system, the issue of *consistency* must be recognized. consistency deals with the issue of the fit between the rate and intensity of quality-related information generated by the manufacturing process and the rate at which the process control system is able to collect, store, summarize and analyze this data.   The lack of consistency becomes a problem when the process control system lags the manufacturing system.   In such situations, the ability of the process control technique to control and monitor quality rapidly deteriorates.   The result is an increase in the levels of scrap, salvage and rework.   Ultimately, quality must deteriorate. Inspection can no longer be avoided.

Consistency, or the degree of *fit* between process control and manufacturing process, is based, in part, on an understanding of the various possible manufacturing environments and the differences that exist between them.   Manufacturing environments are not homogeneous.   As a result, the same techniques for process control should not be applied to all systems.   This paper focuses on the issue of consistency to show that there is a specific type of manufacturing setting in which current body of process control techniques is deficient.   Most process control techniques are intended for use in manufacturing conditions characterized by low to medium levels of information intensity.   Examples of these techniques include 100 percent inspection and Statistical Process Control (SPC).   However, these techniques are severely taxed when applied to high volume, highly automated, information intensive, short cycle manufacturing systems.   such environments are referred to as *Automated-Integrated Manufacturing Systems (AIMS)*.   The speed of data generation combined with the sheer volume of information often overwhelms these conventional procedures.   The AIMS domain requires that the process control technique approximates *real-time* response.

A new approach to process control is needed for this manufacturing setting.   This paper introduces such a procedure, Sufficient Statistic Process Control (SSPC).

The specific objectives of the paper are to:

1. Provide an overview of process control and the conditions that influence the selection and structure of *acceptable* process control techniques.

2. Examine the various manufacturing settings and to identify the specific requirements that they impose on process control systems.
3. Describe the major operating characteristics of the Automated-Integrated Manufacturing System.
4. Investigate the issue of *fit* between manufacturing domains and process control techniques.
5. Develop the conceptual foundations of Sufficient Statistic Process Control.
6. Identify and discuss the major steps in SSPC.
7. Describe alternate applications of SSPC to process control problems.

Because it is able to reduce the volume of data generated by AIMS while maintaining the necessary process quality information contained in that data, SSPC is well suited to the demands of the Automated-Integrated Manufacturing System. As a result, it ensures that for every manufacturing setting, there is now an appropriate supporting process control procedure.

Process Control:
An Overview

Quality control can be thought of as "...the regulatory process through which we measure actual quality performance, compare it with standards, and act on the differences." Juran [1974] Quality control is an umbrella embodying all of the actions needed to establish, maintain and improve overall quality in the firm. Included in this activity are issues of planning for quality, improvement of quality and the control of quality. Wadsworth, Stephens & Godfrey [1986, p. 354] It is a broad based, integrative system which coordinates the activities of other departments such as marketing, manufacturing and engineering in support of a given level of quality.

Process Control Process control is part of the quality control umbrella. Process control is applied at the shop floor level. Its focus is primarily on quality improvement and the control of quality. Wadsworth et al. [1986, Chapter 10] The term process control collectively describes that set of techniques which use information generated from the inspection of a product or service during production to determine what actions, if any, are necessary to maintain quality levels of a given product or process Duncan [1974]. Among the techniques used in process control are control charts, sampling inspection procedures and the statistical design of experiments. AT&T [1985, p. 4]

Process control has two major uses. First, it measures the current quality levels. Second, it helps detect whether the process itself has changed in a way that affects quality. These functions are provided by the five major activities of process control:

8. *Data Generation:* The physical process of capturing the data, through either manual or automatic means (e.g., by sensors) generated by the manufacturing process.

9. *Data Storage:* The permanent accumulation of data (e.g., on paper, magnetic tape, data base records) for use by the subsequent activities.

10. *Data Summarization:* Simple data reduction of the information collected in data storage. Often summarization involves calculating and reporting wither simple descriptive statistics or bar graphs.

11. *Analysis of Data:* The process of monitoring the collected and stored data for potential quality-related problems. analysis attempts to identify those conditions indicative of potential quality control problems (e.g., shifts and trends).

12. *System Reaction:* How the system reacts to flagged problems. Questions such as who is responsible for taking corrective action, under what conditions must corrective actions be taken, and what are acceptable corrective actions must be addressed when carrying out this final activity.

Taken together, these five activities form a closed loop process control system.

Process Characteristics   The manufacturing environment affects the structure of the process control system by determining the conditions under which the control system must operate. The most important of these conditions can be described as follows:

1. *Number of Performance Measures to be Collected:* The first measures refers to the number of measures which

must be collected to properly monitor the quality of either the product or process. In general, the complexity of the process control system is positively correlated with the number of performance measures collected.

2. *Type of Performance Measures Collected*: Performance measures are either value or attribute based. Value measures are ratio or interval (e.g., the size of drill holes or the amount of current passing through a connection) while attribute measure are nominal (e.g., pass/fail). The collection of value measures is more demanding. The complexity of the process control system is greatest in those situation where both value and attributes must be collected.

3. *Speed of Data Generation:* This condition describes the physical rate at which the manufacturing system produces information. In general, the faster the rate, the greater the pressure on the process control system to keep up with the flow of data. The combination of the speed of data generation and the type of performance measures collected define the degree of information intensity.

4. *Degree of System Automation:* Manufacturing systems span the continuum running from low automation to high automation. The extent of system automation is determined by two factors. The first is the type of technology used; the second is the volume (level of output). In general, low automation manufacturing systems are typically labor intensive, low volume processes. In contrast, high automation systems are capital/technology intensive and high volume.

5. *Relative Reaction Time:* Reaction time is the elapsed time interval between the moment that a problem in the process occurs and the moment that corrective actions begin. The relative reaction time is the ratio of the reaction time (T_r) to the per unit process time (T_p). The longer the reaction time relative to the processing time, the higher the level of product loss. While the reaction time is determined by the process process control system, the per unit process time is influenced by the manufacturing process. Relative reaction time (RT_t) can be expressed mathematically as follows:

$$RT_t = T_r/T_p$$

6. *Source of Quality Problems:* At any point in time, a quality problem can be caused by difficulties in either the process, the inputs or the product design (or some combination of these factors).

7. *Identifying the Actual Source of Data Variation:* Variations in quality data can be attributed to one of four possible factors: (i) actual variations in the process; (ii) variations in materials; (iii) instrumentation error (the calibration of the measuring instrument deteriorates); or, (iv) operator error.

Of these seven dimensions, it is the degree of information intensity that is the most critical. The degree of information intensity can be used to differentiate between the process control requirements of different manufacturing settings. Changes in the degree of information intensity present in a manufacturing system often require changes in process control techniques.

Manufacturing Environments and Process Control Techniques

Manufacturing and process control can take place in a wide range of settings. Included are such environments as the project, job shop , batch shop, repetitive manufacturing and process flow. Traditional, these various manufacturing environments have been differentiated using such dimensions as: type of equipment (general purpose versus special purpose); volume; frequency of production (i.e., how often an order for a particular part was requested); order quantities (i.e., typical batch sizes); routings; labor skill levels; and, ability to accommodate changes (i.e. ease of line changeovers and setup times). There is one dimension missing which is critical when evaluating settings from a process control perspective. This is the degree of information intensity.

Using this criterion, we can categorize the various settings. As a result, we have a spectrum running from very low information intensity to very high information intensity. the degree of information intensity determines to a large extent the most appropriate type of process control technique to use.

At the low end of the spectrum, we would find settings such as the project and the job shop. In both cases, the small order quantities (frequently in batches of one or two) are combined with low frequency of production and long processing times. In such situations, the most appropriate technique for process control is that of 100 percent inspection. Because of the low production volumes, long processing times and low frequency, procedures drawing on sampling techniques are not applicable. We have the time to check every piece produced.

The second major category consists of those settings characterized by medium degrees of information intensity. This category begins at one end with the batch shop and can include repetitive manufacturing and, at the other extreme, process flow. Even though production volumes may be high and frequency high, the selection of a limited number of control attributes permits the use of a sampling based procedure such as Statistical Process Control (e.g., X and R charts or cum charts).

The high end of the spectrum is a relatively new development. Here, the following characteristics are found:

* quality-related data is continuously and automatically generated;
* the per unit processing time is short;
* there are multiple measures consisting of both value and attributes;
* the level of system automation is high;
* the process cannot be stopped except when necessary to correct problems; and,
* the volume of process data is large and growing.

In such an environment, the various activities of process control must be done as much as possible in "real time." These various characteristics create a volume of information which is both sufficiently large and rapidly generated as to overwhelm a sampling-based procedure such as Statistical Process Control. As a result, there is a need for a new and different technique for process control which is suitable for use in such a high information intensity environment (see Figure 1).

Such manufacturing characteristics are generally found in Automated-Integrated Manufacturing Systems (AIMS). The AIMS environment is a recent development and is the result of the marriage of advances in both manufacturing processing and information collection technologies.

	Manufacturing Setting	Process Control Technique
Low Info. Intensity	Project	100 % Inspect
	Job	100 % Inspect
Medium	Batch	SPC
	Repetitive/Line	SPC
High Info Intensity	AIMS	?

Figure 1. Representing the Fit between Process and Process Control Techniques

Automated-Integrated Manufacturing Systems: An Overview

As noted previously, the automated-integrated manufacturing system is characterized by the presence of high volume short cycle production processes which are highly integrated, highly automated and very information intensive. To understand the nature of AIMS environment, it is important to understand the foundations on which this system is built -- the developments in manufacturing and information technologies.

The first component, manufacturing technology, consists of such developments as robotics, flexible manufacturing systems and automated guided vehicle systems (AGVS). These developments have changed the way in which material is moved, stored and processed. They have helped management greatly reduce lead time and increase the opportunity for greater product variety. They have also enhanced the overall productivity of the shop floor.

Complementing these advanced have been such developments in information technology as machine vision, improved sensors, automated identification (e.g., bar coding) and standardized communication protocols (e.g., MAP and TOP). These information developments allow the rapid and accurate collection and reporting of data in an online, real time setting. More importantly, these developments have enabled the process of data collection to keep pace with the rate of production. together, these two developments are responsible for the high level of information intensity which is characteristic of the AIMS environment.

Statistical Process Control and the Aims Environment Statistical Process Control (SPC) is not well suited for process control the AIMS domain for three major reasons. First, SPC relies on various control charts (e.g., R charts, p-charts, np-charts, and c-charts) which are generated through sampling. In an X chart, for example, all observations in each sample are reduced to a single measure - the mean. The mean is an imperfect indicator of change for two reasons:

* It masks the time ordering observations.
* The mean lags the actual rate of change taking place.

Any attempts to make these control charts more sensitive to the time ordering of data by reducing the sample size is doomed to failure. The penalty paid for any reductions in the sample size is an increase in the control limits.

Second, SPC requires the storage and analysis of extensive amounts of information. For example, for a simple p-chart, SPC requires that the past means be stored as well as observations needed for the calculations of the current mean. This requirement creates two major problems. First, the data is frequently only summarized with little analysis taking place at the time that the data is stored. Second, there is often a significant time lag between when the data was generated and when it was examined. This lag could be as short as one day or as long as one month. During the lag, the system can be creating *waste*.

In most instances, the costs of using SPC (increased storage requirements and increased lead times) are acceptable. However, in an AIMS settings, the slow response rate is one cost imposed by SPC which makes it prohibitive to use. SPC cannot keep pace with the production requirements of an AIMS domain. It is too slow and it requires the storage of too much information.

An Example A study carried out by a major international electronics corporation based in the United States best illustrates the difficulties of using SPC in an AIMS setting.

The study showed that one of its AIMS lines was routinely producing 2000 circuit boards a day. For each board, an average of 800 measures (attribute and value) were recorded. For one day, the process control system collected, recorded and analyzed some 1,600,000 pieces of information or 6,400,000 bytes of data (at four bytes per measure). In addition, adequate process control required over 10,000 separate analyses per day (conservatively estimated).

Statistical Process Control could not be used in this setting for several reasons. First, the storage requirements needed by SPC were too extensive. Since the data generated was recorded and processed through a data base system, the volume of data quickly consumed most of the computer capabilities each day. Second, SPC caused a significant deterioration in computer response time. Each of the 10,000 analyses required a separate read of the data. Data reads were time consuming since the read requests had to be processed through the data base management system. In addition, data reads and data analysis had to compete with data generation and data recording activities for computer time. Data recording took precedence over data reads and data analysis. The problems with computer system performance increased exponentially as more data were collected. Finally, the relative response time for SPC was too long. by the time that data indicating a problem were recorded and analyzed, too many bad circuit boards had been produced.

Sufficient Statistic Process Control: An Overview

Sufficient statistic process control is designed to monitor a data-intensive, high volume manufacturing process in as close to real time as possible, identifying process operations that are creating defects, and to analyze the data to specify the underlying causal factors. The SSPC procedure is built on several key points:

1. The data from process operations must be intercepted immediately.

2. The analysis of data cannot depend on accessing large data bases. Time is lost when analysis programs have to stop and constantly read in large historical data files. such data bases

must be compressed and summarized by a few statistics that are stored in a smaller file that can be quickly accessed. The resulting statistics are referred to as *sufficient.*

3. Sufficient statistics must be easily computable. For the historical data to be accurately represented, these sufficient statistics must be continuously updated whenever new data is received. This requires the development of recursive formulae that combined the current sufficient statistics with the data that has just been received.

4. Analysis calculations must be preformed in an efficient manner.

5. Reporting of potential problems must be immediate. Often the most immediate method of getting quality related information back to the responsible operating station is through the use of computer generated graphics. In addition, graphics offer another major advantage. Graphical reports can often be interpreted much more rapidly than numerical reports.

The development of a procedure such as SSPC is critical because it ensures the maintenance of a fit between the requirements of the AIMS domain and the process control technique used to maintain quality in that setting (see Figure 2)

The SSPC methodology incorporates the empirical Bayes statistical approach to estimate at any given point in time the

	Manufacturing Setting	Process Control Technique
Low Info. Intensity	Project	100 % Inspect
	Job	100 % Inspect
Medium	Batch	SPC
	Repetitive/Line	SPC
High Info Intensity	AIMS	*** SSPC ***

Figure 2. Process/Process Control Fit Positioning SSPC

distribution of the average response for a process parameter (i.e., $\mu_t \mid X_t$). A process parameter is a measure used to capture the state of the manufacturing process at a given point in time. Process parameters are usually continuous or analog in nature, pertaining, for example, to the electrical or physical measurements of the product manufactured. comparison of the process parameters with the design limits is used to evaluate whether or not the manufacturing process is under control.

Bayes Theorem allows SSPC to deal explicitly with the time dimension of the data -- a dimension ignored by conventional process control techniques. The use of the empirical Bayes Theorem is not unique. Hoadley [1981] and Phadke [1981] used empirical Bayes techniques to address quality problems. However, their procedures were not intended to address quality problems in real time. Hoadley and Phadke used six week time slices applied to a discrete process. SSPC, in contrast, is intended for use with data generated by a continuous process.

The SSPC Procedure Described   The operation of SSPC is conceptually relatively straight forward. Two distributions of the average response for a process parameter are generated. The first distribution represents the long term state of the process while the second distribution denotes the short term condition. Movements of the short term distribution away from the long term distribution indicate the presence of quality control problems in the underlying process. Implementation of SSPC typically consists of six major steps. These steps are summarized in this paper. Any reader interested in understanding the theoretical underpinnings of SSPS is advised to read Sturm, Melnyk, Feltz and Wolter [1989].

Determine the Weights for Short and Long Term Distributions
Of critical importance to SSPC is the weighting factor, p^{T-t}, where T represents the time period for the most current observation and t is the time period for a past observation (such that T-t > 0). This factor, p^{T-t}, is the weight given to each time unit, where p is an arbitrary number usually drawn from the interval (.9,1.0). Specifically, p^{T-t} represents the weight assigned to an observation occurring T-t time periods ago. The p^{T-t} is selected so that it determines at what point past history has essentially a zero weight. The p operates in a manner

analogous to that of the a in exponential smoothing. The effective sample size (N) for any given p value can be calculated as follows.

$$N = 1/(1-p) \qquad (1)$$

The SSPC methodology requires two different values for p. the first value (p_s) is a short term weighting factor while the second (p_l) is the long term weight. These two values require that two different sets of distribution estimates be generated. One set based on p_s tracks short term shifts in the distributions: The other, based on p_l, monitors long term changes.

Experience with SSPC has shown that values for p_l are best selected from the high end of the range [.9,1]. The short term p value, p_s, on the other hand, should be picked from the low end. In applications of SSPC, typical p_s values have come form the interval [.90,.95] while p_l is most often set at .99. Generally, p_l should be between 5 to 20 times larger than p_s. Of the two p values, greater care should be used when choosing p_s.

The short term weighting factor should be selected so that it is consistent with two issues. The first is the beliefs of the users about the stability of the process. the more unstable that the process is perceived, the lower the p_s should be. The second is the cost of estimate error. The higher the cost of estimate error (i.e., the costs incurred in correcting flagged quality control problems), the lower the p_s should be.

At present, both p values are treated as fixed. They are not dynamically modified to reflect changes in the user's beliefs about the process and its stability (a belief based on observations). These values must be selected *before* we can begin the operation of SSPC.

Initialize the Estimates
Next, estimates describing the two distributions must be generated. To do this, an initial pool of data must be created. Enough observations must be collected so that estimates of two variances, σ^2 and τ^2, can be made. The stability of the process determines the length of this initialization period. For a stable system, the initialization period can be as short as 20

observations. As the level of system stability decreases, the length of the initialization period must increase. The primary intent of this initialization period is data collection. Analysis of the data and any needed corrective actions take place after the completion of this stage.

Derive the Sufficient Statistic Estimates

The objective of SSPC is to estimate μ_t given the time series of process parameters (i.e., X_t's or the process parameter values at time t). This requires that the three distribution parameters, μ, σ^2, and τ^2, be estimated. The value, μ, is the mean of the process. The two variances identified describe different types of variability. The first variance, σ^2, is the variation of X_t at time t and contains measurement errors as well as natural fluctuations of the product response when the process is centered at μ_t. The second source of variation, τ^2, is due to the process changing over time. This estimation procedure is based on two assumptions. First, the X_t's are assumed to be normally distributed about the mean μ with variability σ^2. Second, the average process response, μ_t, is assumed to vary over time, with the variations normally distributed with mean μ and process variance, τ^2.

Estimates of these three values, μ, $\sigma^{2'}$, and $\tau^{2'}$, are generated using equations 2-5. all estimates are denoted by '.

$$\mu' = \frac{\Sigma_t p^{T-t} X_t}{\Sigma_t p^{T-t}} \tag{2}$$

$$\sigma^{2'} = \frac{\Sigma_t p^{T-t} (X_t - X_{t-1})^2}{2\Sigma_t p^{T-t}} \tag{3}$$

$$V' = \frac{\Sigma_t p^{T-t} (X_t - \mu')^2}{\Sigma_t p^{T-t}} \tag{4}$$

$$\tau^{2'} = V' - \sigma^{2'} \text{ if } V' - \sigma^{2'} > 0 \tag{5}$$

$$= \delta\sigma^{2'} \text{ otherwise}$$

where δ is an arbitrarily small number ($0 < \delta < 0.1$).

Equation 2 reflects the assumption that the X_t's are ordered in time $(X_1, X_2, ..., X_T)$ so that X_T is the most current observations. Estimating t^2 is more complicated and requires the use of equations 4 and 5. The parameter V' is a first order approximation for V (see Sturn, Melnyk, Feltz and Wolter [1989] for more detail).

Equations (2) to (4) can be rewritten recursively as equations (7) to (9). by using a recursive formulation, we can reduce the amount of information that must be stored from T-t observations to only four points. These recursively derived estimates are called **Sufficient Statistic Process Estimates**.

$$n_T = \Sigma_t^T \, p^{T-t}; \quad n_{T-1} = \Sigma_t^{T-1} \, p^{T-t} - 1 \tag{6}$$

$$\frac{\mu'T = X_T + \mu' \, T - 1 + \mu' \, T - 1}{n_T} \tag{7}$$

$$\sigma'2_T = \frac{n_{T-1} \left(p\sigma' \, 2_{T-1} + \left((X_T - X_{T-1})^2 / 2n_{T-1} \right) \right)}{n_T} \tag{8}$$

$$V' = \frac{n_{T-1} p V'_{T-1} + (X_T - \mu'_{T-1})^{2*} (n_t - 1) / n_t)}{n_T} \tag{9}$$

Estimating the Distribution Parameters

Having established the short and long term distributions in the previous step, the next step is to calculate the distribution parameters. We know that the distribution of μ_t given X_t is a normal distribution with mean (equation 10) and variance (equation 11).

$$w\mu_t + (1-w)X_t \tag{10}$$

$$\tau^2 w \tag{11}$$

where w is defined as:

$$w = \sigma^2 / (\sigma^2 + \tau^2) \tag{12}$$

The parameters required for equations 10-12 (i.e., μ, σ^2 and τ^2) are estimated empirically from the X_t's (as previously described in step 3). These substitutions result in two empirical Bayes' estimates of the distribution $f(\mu_t \mid X_t)$. Again, these two distributions represent short term and long term states.

Develop Adequate Reports to Represent SSPC Results

For SSPC to be effective, its results must be clearly and quickly related to the user. One method of achieving this result is to graphically display the results. Such a graph must display three items of information: (i) the engineering limits or design specifications; (ii) the long term distribution, $f(\mu_t \mid X_t)$; and, (iii) the short term distribution. Representing the engineering limits is straight forward.

A box graph is an effective means of displaying the long term distribution. A box graph offers several attractive features. Chosen percentiles can be compared effectively and quickly. Cleveland [1985, p. 132] They can also convey a great deal of information visually. to draw a box graph, we need to determine the normal percentile. This process can be done as follows.

From the third step, we derived the estimates for μ, σ^2 and τ^2 by the use of recursive formulae (6) - (9) and substitution in equations (10) - (12). Using the weight for the long term distribution, we produced an Empirical Bayes estimate of the long term distribution. Using this estimate in conjunction with the algorithm developed by Kennedy and Gentle [1980], we can now calculate normal percentiles. A box plot can now be drawn.

The same procedure is repeated for the short term distribution. However, drawing both short and long term box graphs on the same display is potentially confusing. It may be difficult for the user to quickly separate out these two graphs. To eliminate this confusion, the display should be simplified by displaying the box graph for the long term distribution in its entirety (since it is the major reference point) and limiting the short term distribution of the mean alone. Both distributions are still calculated and tracked. The user should be warned by a note on the display if the difference between the two distributions is significant (the procedures for determining when differences are significant are discussed later). an example of the resulting graphic display is shown in Figure 3.

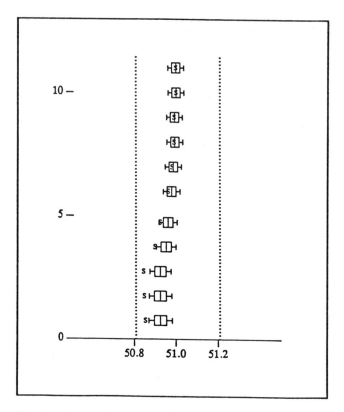

Figure 3. An Example of an SSPC Graphic Display

Periodically Review the SSPC Procedure
Periodically, the data base should be purged. This purging is required to update the data base to account for product phase ins and phase outs. In addition, the p values should be reviewed to determine if they should change to reflects new conditions in the process. this revision should be done in consultation with the users. In practice, purging can be done automatically and as frequently as once a day.

Identification of Process Control Problems Most process control problems encountered fall into one of three categories:

* the process is no longer within engineering design limits;

* the process, through sudden shifts, is moving out of process control; and,

* the process control problems are due to problems with calibration and not to problems with the underlying process.

Exceeding Design Limits   Dealing with the first category is straightforward.   Whenever any portion of either distribution crosses the limits, the process is no longer within engineering specifications.   In most cases, movements in the short term distribution will indicate the presence of a problem. Statistically, the point at which distributions exceed the design limits can be determined as follows:

For a process to pass engineering inspection, X_t must be between the engineering limits, L_1 and L_2. Equation 13 measures how much of the process distribution is within the engineering limits.   This equation gauges the amount of the process distribution falling between the design limits.

$$P_I = \text{Prob}\left(\mu_t \, \varepsilon \, (L_1, L_2) \,|X_t\right) = \int_{L_1}^{L_2} f(\mu_t \,|\, X_t) \, dUT \qquad (13)$$

This statistic can be computed by standardizing the integral and using equation 14, as derived by Hastings [1955].   this equation approximates the cumulative probability of a standard distribution up to the point z.

$$P_1 = .353553 \; \{1-[1/(1+.278393z+.230389z^2+.000972z^3+$$

$$.0781808z^4)^4]\} \qquad (14)$$

If the process is running smoothly, P_1 should be close to one (i.e., all of the distribution falls within the limits).   If P_1 drops below a prespecified limit (e.g., .95 or .90), a change has been detected, indicating that the process is beginning to drift.

Sudden Shifts   Sudden shifts in the process occur whenever the short term distribution pulls away from the long term distribution.   The graphical reports can help the user judge if such drifting has occurred.   To determine statistically if such a sudden shift has occurred requires testing the difference in distribution means.

To test the difference in distribution means, we must first calculate z, or:

$$z = \frac{\mu_{lt} - \mu_{st}}{\sqrt{(var_{lt} + var_{st})}} \qquad (15)$$

Where μ_{lt} and μ_{st} and var_{lt} and var_{st} are the respective mean and variance of the long and short term distribution as calculated from equations 10-12.

Next, P_2 is calculated by applying equation (14). The value P_2 represents the probability of the short term process mean being different from the long term process mean. In the absence of a sudden shift, P_2 will be close to 1/2. A sudden shift is indicated whenever P_2 shifts towards either 0 or 1. In practice, the limits on P_2 are empirically set to correspond to the amount of drift tolerable.

<u>Instrumentation Problems</u>   Identifying hardware of calibration problems requires the presence of multiple *identical* test sets. One approach is to do pairwise comparisons of the test sets. For any number of test sets greater than 2, this procedure is very inefficient and time-consuming. To cope with such situations in real time, a new efficient test statistic is derived in equation (16). this test statistic, G, will detect test set differences when measuring the performance of a process parameter.

$$G^2 = \Sigma^n_{i=1}(\varepsilon - \varepsilon_\iota)^2/(\tau^2 w_i) \qquad (16)$$

Where n is the total number of test sets, i denotes the ith test set, e_i is the empirical Bayes estimate of the mean (equation 10) of the ith test set, $\tau_i^2 w_i$ is the empirical Bayes estimate of the variance of the ith test set (equation 11) and e is the average of the e_i's. G^2 has an approximate X^2 distribution with 1 degree of freedom.

Sufficient Statistic Process Control: An Example

To illustrate the operation of SSPC, a hypothetical example of a circuit board assembly and test line is used. In this shop, which satisfies the conditions for automated integrated manufacturing system, the volume of production is routed through one of five identical test sites. Each circuit board must be tested at one of these sites before it can leave the system. Each test site has a 51 volt power supply for powering up the boards. The level of these power supplies is monitored continuously as each board is tested. Engineering tolerances for these power supplies are 51 volts ± .2 volts. The voltage levels must remain inside these bounds.

SSPC is used to monitor these output voltages (the Xt's over time) and to detect when these voltages begin to get close to the bounds or when the voltages suddenly shift, indicating a change in the underlying process. Under SSPC, the two distributions are updated with each measurement. As a result, the distributions are continuously changing over time. To understand how SSPC identifies quality problems, consider one of the test sets, TS30.

An example of the recordings made at TS30 for a 10 minute period is given in Table 1. It is assumed that the prior to this period, the process was stable for some time. The short term distribution is based on ps = .95 (N = 100 observations). The box graph for the long term distribution uses p1 = .99 (n = 100 observations). The box graph for the long term distribution explicitly identifies the 5th, 10th, 50th, 90th and 95th percentiles. To simplify the example, only the mean of the short term distribution is monitored.

There are no quality problems for the first 4 minutes. The results for minute 5 begin to indicate a slight drifting, as shown by the movement of the short term mean, s, to the left of the long term mean. the process is still under control. By minute 6, the short term mean has moved to the extreme left of the box graph. A trend is being indicated. At minute 7, the short term mean has moved outside the box graph and a quality problem in need of correction has been flagged. The items produced at minute 7 are still acceptable (they fall within the control bounds). however, the movement to the lower limit shown in the process control tests indicates that unless the problem is corrected, the process will eventually produce unacceptable items. correction now will ensure that all boards produced are still acceptable.

Table 1   SSPC in Operation (An Example)

Time	Obs	Short Term-Dist of $\mu_t \mid X_t$ N(Mean, Var), $\hat{\mu}, \hat{\sigma}^2, \hat{\gamma}^2$	Long Term Dist of $\mu_t \mid X_t$ N(Mean, Var), $\hat{\mu}, \hat{\sigma}^2, \hat{\gamma}^2$	Box Plots
0		51.0000,.0008,51.0000,.0057,.0009	51.0000,.0008,51.0000,.0060,.0010	H[§]H
1	51.01	51.0019,.0008,51.0005,.0054,.0009	51.0015,.0008,51.0001,.0059,.0010	H[§]H
2	50.96	50.9929,.0007,50.9985,.0052,.0009	50.9940,.0008,50.9997,.0059,.0010	H[§]H
3	50.97	50.9930,.0007,50.9971,.0050,.0009	50.9952,.0008,50.9994,.0058,.0010	H[§]H
4	50.93	50.9826,.0008,50.9937,.0048,.0010	50.9885,.0009,50.9987,.0058,.0010	H[§]H
5	50.89	50.9651,.0011,50.9885,.0046,.0014	50.9801,.0009,50.9976,.0057,.0011	H[§]H
6	50.83	50.9268,.0016,50.9806,.0044,.0025	50.9639,.0011,50.9960,.0057,.0014	B[ ]H
7	50.82	50.9026,.0019,50.9726,.0042,.0036	50.9546,.0013,50.9942,.0056,.0016	S[ ]H
8	50.76	50.8465,.0023,50.9619,.0040,.0054	50.9270,.0016,50.9918,.0056,.0022	S H[ ]H
9	50.79	50.8509,.0024,50.9533,.0039,.0065	50.9268,.0017,50.9898,.0055,.0025	S H[ ]H
10	50.81	50.8564,.0024,50.9462,.0037,.0072	50.9272,.0019,50.9880,.0055,.0028	S H[ ]H

50.8   51.0   51.2

SSPC:
Limitation   and   Cautions

Present in SSPC are certain limitations.   The following are the most critical:

a. <u>SSPC is most appropriate for high volume manufacturing settings:</u>   Because the Bayesian approach was designed for high volume manufacturing, it is assumed that the items tested in sequence have seen basically the same processing. This assumption assures that the corresponding measures come from basically the same distribution.

b. <u>Detailed tracking and recording of past historical quality data is difficult under SSPC:</u>   Because of the data reduction approach present in SSPC, the user cannot go back to time t-1.

c. Stability of correlations: SSPC assumes that the process location parameters (μ_t's) remain relatively stable over small time intervals.

d. Initial stability: SSPC requires that the process being controlled have a period of initial stability so that the model parameters being estimated by SSPC have time to converge statistically.

e. Normal distribution: SSPC assumes that the process being controlled is normally distributed.

For a more complete discussion of the limitations and cautions applicable to SSPC, the reader is referred to Sturm, Melnyk, Feltz and Wolter [1989].

Conclusions

An effective process control system is an essential component of any company's quality control program.   For the process control system to be effective, the techniques it uses must be consistent with the operating conditions and quality control requirements of the specific manufacturing environment.   This paper has shown that there exists an area in which current process control techniques do not fit well with the requirements of a new type of manufacturing environment. This is the automated-integrated manufacturing system.   This is an environment characterized by an extremely high level of information intensity.   In such a setting, procedures such as either 100 percent inspection manual or by machine) and Statistical Process Control are not appropriate.   They are quickly overwhelmed by the large amount of information quickly created by this system.   As a result, a new and different technique for process control is required.

This paper introduced such a technique, that of *Sufficient Statistic Process Control* (SSPC). This is a procedure which is able, by means of the extensive use of Bayesian theorems and techniques, to reduce the vast amount of information to a few items (i.e., *Sufficient Statistics*) of information which are then continuously and easily updated. The result is the reestablishment of the fit between the process and the process control technique.

The concept of fit introduced in this paper is critical for it demands that both researchers and managers be continuously aware of the delicate balance that exists between manufacturing process and process control technique. As changes in the manufacturing process take place, we must be ready to assess these changes to determine what effect, if any, the new manufacturing process will have on the resulting structure of the process control system. If the structure of the process control system must be changed, then these changes must be introduced quickly. the goal is to maintain the fit.

Failure to maintain this fit will result in a situation where the process control system is no longer an asset to the manufacturing process, but rather a hindrance. when this occurs, quality is no longer necessarily inherent in the process. The costs of such a situation are borne by the entire manufacturing and corporate system.

References

AT&T Technologies. (1985) Statistical Quality Control Handbook. Second Edition Eleventh Printing, Charlotte, NC: Delmar Printing Company.

Cleveland, W.S. (1985) The Elements of Graphic Data. Monterey, CA: Wadsworth Advanced Books and Software.

Duncan, A.J. (1974) Quality Control and Industrial Statistics, Fourth Edition. Homewood, IL: Richard D. Irwin.

Hastings, C. (1955) Approximations for Digital Computers. Princeton, NJ: Princeton Press.

Hoadley, A.B. (1981) "The Quality Measurement Plan." Bell System Technical Journal, pp. 215-217.

Juran, J.M., Editor. (1974) Quality Control Handbook. Third Edition, New York, NY: McGraw-Hill.

Kennedy, W.J. and Gentle, J.E. (1980) Statistical Computing. New York, NY: Marcel Dekker.

Phadke, M.S. (1981) "Quality Audit Using Adaptive Kalman Filtering." ASQC Quality Congress Transactions. San Francisco, pp. 1045-1052.

Sturm, G.W., Melnyk, S.A., Feltz, C.J. and Wolter, J.F. (1989) "Sufficient Statistic Process Control: An Empirical Bayes Approach to Process Control." Working Paper, Department of Mathematics, Grand Valley State University, January.

Wadsworth, H.M., Stephens, K.S. and Godfrey, A.B. (1986) Modern Methods for Quality Control and Improvement. New York, NY: John Wiley and Sons.

5
MONITORING THE EFFECTIVENESS OF REFINERY ADVANCED CONTROLS WITH A STATISTICAL PROCESS CONTROL SCHEME

Catherine M. Jablonsky
Staff Associate
Luftig and Associates, Inc.
Farmington Hills, MI 48018

John R. Davis, P.E.
Senior Control Engineer
Profimatics, Inc.
Thousand Oaks, CA 91361-1072

Gregory Hitchings
Senior Process Engineer
Ultramar
Long Beach, CA 90801-0920.

THeRe ARe SPC
To APPLICABLE
To oTHeR INDUSTRIes
oTHeR THAN

Abstract

The application of statistical process control to discrete processes is well documented and proven to be of great value. However this is not the case with continuous processes; i.e., processes producing a product that cannot be divided into meaningful discrete units. This paper documents a case study of the use of statistical process controls to monitor, in an automated fashion, the dynamic response of a continuous process, in order to discover out of control conditions. These conditions are then evaluated to determine the cause.

The application described is for automatic monitoring of the advanced cutpoint controls installed on two crude units in an oil refinery. The statistical technique used here is a Shewhart-CUSUM control scheme. Problems applying this statistical control scheme, selection of the needed Shewhart-CUSUM parameters, data collection and analysis, application of these results for construction of Pareto charts, and possible reaction plans are discussed.

Introduction

This paper discusses an application of statistical process control to a continuous process. Although the statistics involved are relatively simple and well known, the unique aspect of this research is found in the application. The exciting aspect of the research is that it provides an easily implemented automated monitoring scheme for complex process controls which can be used to not only detect degradation in performance, but also assist in determining the cause.

In this paper, a continuous process is defined as a process producing a product which cannot be divided into meaningful discrete units. Advanced controls are defined as control strategies that use multiple process measurements to indirectly estimate and control important product properties.

The monitoring of advanced process control schemes implemented on continuous processes is complicated by both the normal variation in the process and the dynamic interactions between the process control loops. The purpose of this research was to develop a statistical process control technique to monitor, in a completely automated fashion, the dynamic response of a continuous process to an advanced control scheme, in order to discover out of control conditions. These conditions would then be evaluated to determine their cause.

The results of these efforts for the monitoring of advanced cutpoint controls installed on two oil refinery crude units indicate that such a monitoring system can determine out of control conditions reliably and be programmed to evaluate the probable cause. Problems applying this statistical control scheme, such asselection of the needed Shewhart-CUSUM parameters, data collection and analysis, application of these results for construction of Pareto charts, and possible reaction plans are discussed below.

Process   Application

This research was conducted at Union Pacific Resources Company's Wilmington, California oil refinery. The processes statistically monitored were two crude units, and the controls monitored were the overhead naphtha cutpoint advanced controls. The data presented here were collected from May through September 1988.

The naphtha cutpoint advanced controls were chosen because they are a relatively simple advanced control application that continuously estimates and controls an important product specification. Their effectiveness in controlling the process at the desired naphtha cutpoint has

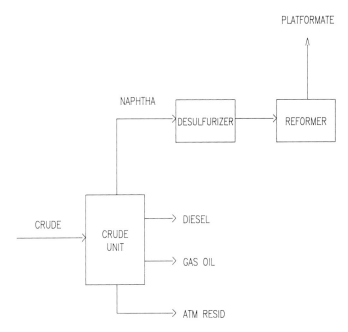

Figure 1   Refinery Configuration

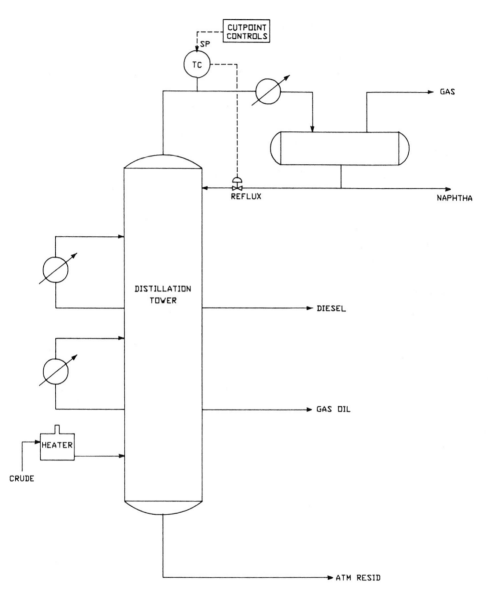

Figure 2(a).   Naptha Cutpoint Control Strategy

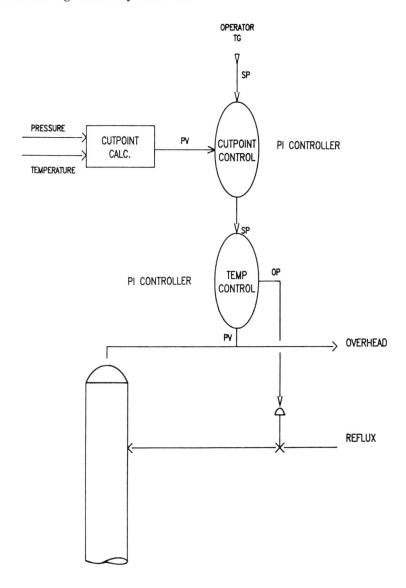

Figure 2(b). Naptha Cutpoint Control Strategy

substantial economic impact. In this case, manually monitoring these controls has proved to be a difficult and time consuming effort.

The portion of the refinery configuration of interest here is shown in Figure 1. Raw crude oil is fed to the crude units which, by distillation, separate the crude into several intermediate products that are then further processed by other refinery units. The overhead product is naphtha that is first desulfurized and then sent to the reformer for upgrading to reformate, a component of gasoline. One of the important reformate product specifications is the boiling point of the heaviest material, commonly referred to as the endpoint. The reformate endpoint is directly effected by the cutpoint of the naphtha produced by the crude unit.

One of the advanced controls implemented on the crude units is the naphtha cutpoint control. The control strategy is shown in Figures 2a and 2b. Temperature and pressure measurements are continuously made and are used to estimate the naphtha cutpoint. This estimated cutpoint is compared to an operator entered target, or setpoint, in a proportional integral feedback control equation (Pervy and Chilton [1973], p. 22-17) (1), which calculates a setpoint for the overhead temperature controller. This setpoint is compared to the measured overhead temperature in another PI control equation, and the reflux valve position is calculated. The reflux valve position controls the amount of reflux (cool liquid) returning to the top of the distillation column and controls the overhead temperature.

The PI control equation used in a closed loop feedback control system is as follows:

$$\Delta OP = K * ((E_N - E_{N-1}) + \frac{\Delta t}{K1} * E_N)$$

(1)

where:

DOP = controller output change

K = proportional gain tuning constant

E_N = (PV - SP), referred to as control error

PV = measured process value

SP = setpoint or target

Dt = time between control calculations

K1 = integral gain tuning constant

This equation uses past and present values of the control error to calculate the change in output. In this case, as seen above, the control error is defined as the difference between the measured process value and the desired value, also known as the setpoint or target. It is important to note that the control errors are not residuals and are not randomly distributed, independent values.

Selection of Shewhart-CUSUM Control Scheme

The parameter chosen for statistical monitoring was the control error because it is directly related to the performance of the controls. The smaller the value of the control error (that is, the smaller the difference between the measured process value and the desired target), the better the performance of the controls. The statistical tool used is a Shewhart-CUSUM control scheme (Lucas [1982]).

A Shewhart-CUSUM control scheme was selected because it could be implemented without the use of graphs, it was simple to program, and it effectively monitors for both rapidly occurring out of control conditions and for gradually developing trends and runs. Its limitations are that it cannot tolerate autocorrelation in the data, and it assumes the measured values of the parameter controlled fits a normal distribution. Also, without the aid of a graph, it may take a significant amount of time to detect a cyclic instability.

The formulas necessary to implement a Shewhart-CUSUM statistical control scheme are simple and easily programmed. Given that the monitored observations are normally distributed, each observation is first standardized using the following equation:

$$Z=(Yi - m)/sigma \qquad (2)$$

where Y_i is the ith observation of the parameter controlled (in this case the control error), m is the target mean value and sigma is the standard deviation of Y. This transforms the data into a standardized distribution with a mean value of zero and a sigma equal to one. Deviations from target in excess of a given reference value, k, are then accumulated into the sums given below.

POSITIVE SHIFTS:
$$Sh_i = MAX (0, (Z_i-k) + Sh(i-1)) \qquad\qquad (3)$$

NEGATIVE SHIFTS:
$$Sl_i = MAX (0, (-Z_i-k) + Sl(i-1)) \qquad\qquad (4)$$

If Sh_i or $Sl_i > h$, the control errors are out of statistical control.

These sums serve to detect positive and negative shifts away from the target mean. If either of these sums exceed the decision interval value, h, an out of control signal is given.

In order to increase the ability of the control scheme to detect relatively large, rapidly occurring shifts off target, Shewhart control limits (SCL) are also added to the control scheme. If any individual standardized value exceeds these limits, i.e., if $|Z_i| >= SCL$, an out of control signal is also given.

When an out of control signal occurs, the CUSUM resets itself to its initial value, So. So may be set equal to zero or some positive value. When a nonzero value is used the control scheme is referred to as a CUSUM with Fast Initial Response. So values greater than 0 are used when it is difficult to reestablish control, and rapid detection of a continuing out of control situation is desired.

The usefulness of the Shewhart-CUSUM control scheme is severely limited unless it is accompanied by some reaction plan. For our purposes, an out of control condition signals the computer to print out a variety of process values which may be the source of the problem. This information was then used to develop a Pareto diagram which prioritized by frequency the causes of the out of control conditions. Once developed, this was used to prioritize process improvements and to establish reaction plans.

Application of the Shewhart-CUSUM Statistical Control Scheme

Data to be statistically monitored cannot be autocorrelated. Autocorrelation exists between successive observations of the control error because the closed loop feedback control system makes the future error a function of the current error. There are two solutions to this problem. One is to fit a predictor equation to the data and subtract the prediction from the actual value obtained to produce a true residual. This residual can then be statistically monitored (See Montgomery and Friedman [1988]). A second, perhaps simpler approach is to extend the time between sampling to a point where autocorrelation does not occur. To determine the validity of this approach, minute by minute values of the control error were fit to an equation using Box and Jenkins time series analysis. Then intermediate observations were eliminated until no autocorrelation was detected. Using the autocorrelation check for white noise provided by SAS it was found that observations one half hour apart were not autocorrelated (see Table 1).

To test for normality of the data when the process was stable, 96 data points were randomly selected and skewness and kurtosis calculations were performed. At an alpha risk of 1%, the data was found to be symmetric but peaked. The results of these calculations are given in Table 2. This result was anticipated given a well tuned controller. That is, the process value was not allowed to deviate significantly from the setpoint via constant process adjustments. Rather than

Table 1 SAS Output

Autocorrelation Check For White Noise

TO LAG	CHI SQ.	DF	PROB	AUTOCORRELATIONS					
6	10.21	6	0.116	0.245	0.054	0.115	0.048	-0.15	0.021
12	13.73	12	0.318	-0.022	-0.120	-0.049	-0.122	-0.021	0.000
18	19.46	18	0.364	-0.001	-0.036	0.107	0.157	0.15	-0.104
24	21.91	24	0.585	-0.105	0.011	-0.087	0.033	0.004	0.008

Table 2
Results of Skewness and Kurtosis Calcualations

Skewness = 0.294   Not Significant at a = 0.01
(Normal = 0)

Kurtosis = 3.372   Significant at a = 0.01

transforming the data to obtain a normal distribution, however, it was decided to assume normality. Given that this assumption would tend to downplay the significance of a deviation off target, it would serve to minimize the possibility of false out of control signals, while keeping the computations as simple as possible. This, of course, increases the risk of missing some out of control conditions. At this stage of the research, however, that was not a significant concern.

For each of the two crude units to be monitored, values of the mean and standard deviation were estimated using historical data. These point estimates are given in Table 3.

Note that the mean values should be equal to zero. Given a confidence interval calculation around the estimate, however, the true means may in fact be zero.

The parameters necessary to implement the Shewhart-CUSUM were selected using the guidelines given by Lucas (p.54). Given the desire to detect a one sigma shift quickly, a k value of 0.5 was selected. This value combined with an h value of 5.2 and a Shewhart control limit (SCL) of 3.5 yields an in control average run length (ARL) of 476 and an out of control, one sigma shift ARL of 10.6 (from Table 3, Lucas, pg.57). Using the fast initial response (FIR) feature, with So = h/2, serves to slightly decrease the in control ARL, however it also serves to

Table 3
Values of the Estimated Means and Standard Deviations

Crude Unit 10   Crude Unit 11
m = 0.2   m = 0.1
s = 3.0   s = 3.3

significantly decrease the out of control ARL when beginning the CUSUM in an out of control state.

Plant Data Collection

Upon implementation of the Shewhart-CUSUM, daily reports of the control error observations monitored and the out of control conditions observed were generated, as shown in Figure 3. These reports also included the time of the observation, the normalized (or standardized) value of the control errors, and the CUSUM values in both the positive (or hi) and negative (or lo) directions. A message column was also provided which flagged out of control conditions and/or when the process was not in computer control.

Data Analysis

To determine the effectiveness of the statistical control scheme, a number of checks were conducted. First, the control errors monitored were graphically displayed and out of control conditions flagged by the Shewhart-CUSUM were verified on the plots. Referring to Figure 3, note the out of control conditions flagged at 13:30 (a shift on the high side of target) and at 20:00 (a shift on the low side of target). The corresponding graph (Figure 4) clearly shows these shifts.

Similarly Figure 5 indicates two out of control conditions. Both of these, however, are cases of the Shewhart control limits being exceeded, in the negative and positive directions, respectively. Again, the corresponding graph (Figure 6) clearly reveals the occurrence of these spikes.

As previously noted, one of the deficiencies of the Shewhart- CUSUM is its potential to overlook the occurrence of a cycle in the monitored observations. Another perhaps greater concern is that by only sampling every half hour, we may be missing some significant peaks embedded in such a cycle. This problem could be addressed by the following methods.

Referring to the bottom of Figure 3, the mean value and standard deviation of both the minute by minute values and the half hour spaced values of the control errors are calculated and printed each day. If peaks and/or cyclic swings are being

<Data printed oldest to newest>

TIME	ERROR	NORMALIZED VALUE	HI	LO
06:30	8.29	2.70	6.99	0.00
07:00	8.29	2.70	4.80	0.00
07:30	8.29	2.70	6.99	0.00
08:00	8.29	2.70	4.80	0.00
08:30	4.42	1.41	5.70	0.00
09:00	-0.98	-0.39	1.71	2.49
09:30	4.30	1.37	2.57	0.63
10:00	4.47	1.42	3.50	0.00
10:30	3.35	1.05	4.05	0.00
11:00	1.82	0.54	4.09	0.00
11:30	0.50	0.10	3.69	0.00
12:00	3.07	0.96	4.15	0.00
12:30	3.54	1.11	4.76	0.00
13:00	2.19	0.66	4.92	0.00
13:30	3.81	1.20	5.63	1.50
14:00	2.00	0.60	2.70	0.60
14:30	1.39	0.40	2.60	0.00
15:00	1.04	0.28	2.38	0.00
15:30	-0.08	-0.09	1.78	0.00
16:00	-0.92	-0.24	1.52	0.00
16:30	-0.23	-0.14	0.88	0.00
17:00	-0.80	-0.33	0.05	0.00
17:30	-0.88	-0.36	0.00	0.00
18:00	0.95	0.25	0.00	0.00
18:30	0.28	0.03	0.00	0.00
19:00	-1.99	-0.73	0.00	0.23
19:30	-9.79	-3.33	0.00	3.06

Time				
20:00	-7.86	-2.69	0.00	5.25
20:30	-5.08	-1.76	0.34	3.86
21:00	-2.46	-0.89	0.00	4.25
21:30	-1.65	-0.62	0.00	4.36
22:00	-2.63	-0.94	0.00	4.81
22:30	-1.07	-0.42	0.00	4.73
23:00	-1.47	-0.56	0.00	4.78
23:30	-1.32	-0.51	0.00	4.79
00:00	-0.73	-0.31	0.00	4.60
00:30	-0.80	-0.33	0.00	4.44
01:00	-0.91	-0.37	0.00	4.31
01:30	-1.05	-0.42	0.00	4.22
02:00	-0.82	-0.34	0.00	4.06
02:30	0.14	-0.02	0.00	3.58
03:00	1.01	0.27	0.00	2.81
03:30	-1.13	-0.44	0.00	2.76
04:00	1.42	0.41	0.00	1.85
04:30	1.22	0.34	0.00	1.01
05:00	9.07	2.96	2.46	0.00
05:30	6.27	2.02	3.98	0.00
06:00	2.12	0.64	4.12	0.00

	MEAN	SIGMA
For last 1440 minutes	1.05	4.01
For last 48 samples	1.02	3.84

Figure 3. Daily Report Sheet--Unit 10

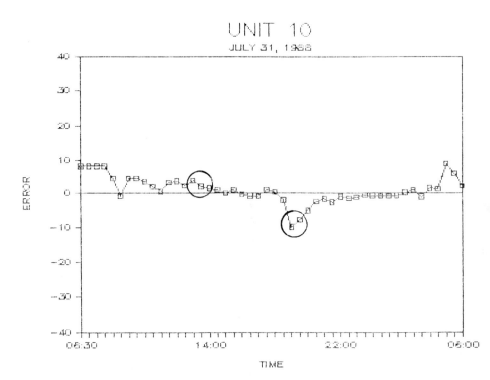

Figure 4.   Error vs. Time from Daily Report Sheet
             (See Figure 3)

CUSUM REPORT -- UNIT 11 CUTPOINT

<Data printed oldest to newest>

TIME	ERROR	NORMALIZED VALUE	HI	LO
06:30	-5.04	-1.56	0.00	0.00
07:00	-4.29	-1.33	0.00	0.00
07:30	-3.16	-0.99	0.00	0.00
08:00	-3.21	-1.00	0.00	0.00
08:30	-1.01	-0.34	0.00	0.00
09:00	-4.28	-1.33	0.00	0.00
09:30	-1.44	-0.47	0.00	0.00
10:00	-0.63	-0.22	0.00	0.00
10:30	-0.24	-0.10	0.00	0.00
11:00	-2.02	-0.64	0.00	0.00
11:30	2.06	0.59	0.00	0.00
12:00	-3.93	-1.22	0.00	0.00
12:30	-5.96	-1.84	0.00	0.00
13:00	-12.50	-3.82	4.92	0.00
13:30	-2.33	-0.74	0.00	0.00
14:00	22.96	6.93	2.70	1.50
14:30	8.08	2.42	0.00	0.00
15:00	1.93	0.55	0.00	0.00
15:30	2.16	0.63	0.00	0.00
16:00	-5.13	-1.59	0.00	0.00
16:30	-3.38	-1.06	0.00	0.00
17:00	-2.19	-0.69	0.00	0.00
17:30	-2.25	-0.71	0.00	0.00
18:00	-1.25	-0.41	0.00	0.00
18:30	1.31	0.37	0.00	0.00
19:00	1.77	0.51	0.00	0.00
19:30	1.45	0.41	0.00	0.00
20:00	2.10	0.61	0.00	0.00
20:30	0.18	0.02	0.00	0.00
21:00	0.41	0.09	0.00	0.00
21:30	1.67	0.48	0.00	0.00
22:00	0.10	0.00	0.00	0.00
22:30	-0.29	-0.12	0.00	0.00
23:00	0.39	0.09	0.00	0.00
23:30	0.32	0.07	0.00	0.00
00:00	-0.34	-0.13	0.00	0.00
00:30	-1.19	-0.39	0.00	0.00
01:00	-0.20	-0.09	0.00	0.00
01:30	-0.46	-0.17	0.00	0.00
02:00	0.89	0.24	0.00	0.00
02:30	0.01	-0.03	0.00	0.00
03:00	-0.50	-0.18	0.00	0.00
03:30	0.46	0.11	0.00	0.00
04:00	-0.03	-0.04	0.00	0.00
04:30	-0.59	-0.21	0.00	0.00
05:00	-0.91	-0.31	0.00	0.00
05:30	0.36	0.08	0.00	0.00
06:00	1.80	0.51	0.00	0.00

	MEAN	SIGMA
For last 1440 minutes	-0.65	4.50
For last 48 samples	-0.38	4.55

Figure 5. Daily Report Sheet--Unit 11

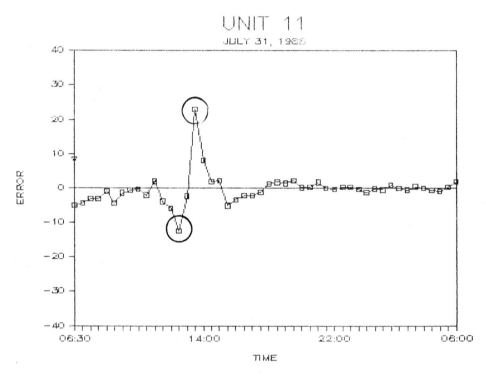

Figure 6.   Error vs. Time from Daily Report Sheet
             (See Figure 5)

overlooked by the CUSUM and sampling technique, the standard deviations and/or means calculated using all 1440 data points versus those calculated using only 48 should differ significantly. Such differences, however, rarely occurred. As illustrated by Figure 7, on only one day between March 31 and May 31 did the standard deviations differ significantly for Unit 10. Similar results were obtained for the standard deviations for Unit 11 and the means for both.

Also, a CUSUM was run on the natural log of the daily variances. Taking the natural log of the variances normalizes the data, as indicated by the tests for skewness and kurtosis for a sample of size of 137 and an alpha risk of 1%. The results are listed in Table 4. Using the transformed data, no out of control conditions were found, indicating no gradual increase or decrease in the variability of the control errors.

Finally, sampling at different intervals (but insuring at least half hour spacing) was used in conjunction with the regularly spaced sampling. No significant improvement in sensitivity to out of control conditions was noted. In light of this result, and since it was much simpler, sampling at regular intervals was the selected method.

As previously mentioned, whenever an out of control condition was noted, the computer was directed to search the

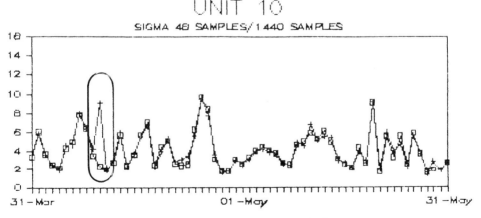

Figure 7. Standard Deviation -- 48 vs 1440 Samples

Table 4   CUSUM on ln(Daily Variances)

CUSUM ON ln σ^2:

<table>
<tr><td>U10</td><td>U11</td></tr>
<tr><td>SKEWNESS = .088</td><td>SKEWNESS = -.157</td></tr>
<tr><td>KURTOSIS = -.701</td><td>KURTOSIS = -.054</td></tr>
</table>

∴ ln σ^2 FOR BOTH UNITS
IS NORMALLY DISTRIBUTED

μ = 2.23	μ = 2.78
σ = 1.36	σ = 1.18

$$h = 5.2$$
$$k = 0.5$$
$$SCL = 3.5$$
$$S_0 = h/2$$

data highway to determine the values of several key process parameters which may be related to the out of control condition. These process parameters were selected by the process control engineers. An example of this printout listing each process value as a 12 minute average at the time of the out of control condition, and at several preceding 12 minute intervals, is given in Figure 8. Using these reports, the engineers were able to verify their suspicions, and Pareto charts for both Unit 10 and Unit 11 were constructed revealing the major causes of the out of control conditions (see Figures 9 and 10). Note that only 12% for Unit 10 and 18% for Unit 11 of the out of control conditions could not be explained by one or more of the process parameters investigated. To explain these unknowns, additional parameter values will have to be added to the data search list. Given that the percentage is relatively small, however, this was not pursued at this time.

The parameters checked were as follows:

1. Valve position for the fuel gas: If the valve is fully opened (greater than 90%) or fully closed (0%), then the heat input to the tower cannot be controlled.

2. Heater setpoint (SP) and heater process value (PV): If the difference between the two values is greater than four or five degrees, then the heat input to the tower is not controlled.

3. Reflux flow rate: If this value is too great (7000 bpd for Unit 10 and 10,000 bpd for Unit 11), then the tower is heat removal limited.

4. Overhead pressure setpoint and process values: If the difference between the two values is greater than one psig, this is an indication of loss of pressure control.

5. Overhead temperature setpoint and process value: If the difference between these two values is greater than four degrees, then effective control of the overhead temperature has been lost.

6. Calculated percent tray 1 flooding was also recorded to see if the tower capacity had been exceeded.

UNIT 11 CUTPOINT
"OUT OF CONTROL"
REPORT <VALUES: L TO R, FROM INDICATED TIME TO BACK IN TIME>

13:00:00

11TIC101	717.7	715.7	715.9	717.9	718.5
11TF101	717.0	717.0	717.0	717.0	717.0
11FV112	76.1	78.5	76.3	73.3	75.3
11CB023	108.0	107.7	107.4	106.9	106.2
11FI109	8.9	8.8	8.7	8.7	8.6
11PIC103	38.0	38.0	38.0	37.9	37.9
11PF103	38.0	38.0	38.0	38.0	38.0
11TIC104	322.5	322.8	323.1	323.3	323.4
11TF104	316.0	316.4	316.3	316.8	316.7

14:00:00

11TIC101	699.7	699.0	697.2	696.8	696.8
11TF101	715.0	715.0	715.0	715.0	715.0
11FV112	75.1	74.1	73.9	74.1	74.4
11CB023	96.2	97.1	97.7	98.9	100.1
11FI109	7.4	7.6	7.7	8.0	8.0
11PIC103	38.9	38.9	39.2	39.1	39.2
11PF103	39.0	39.0	39.0	39.0	39.0
11TIC104	300.6	300.5	300.6	300.9	301.3
11TF104	316.9	316.5	316.4	316.0	316.0

Figure 8. Sample Printout --12 Minute Average

716. 8	715. 4	715. 7	717. 2	717. 9	717. 4	715. 4
717. 0	717. 0	717. 0	717. 0	717. 0	717. 0	717. 0
78. 8	79. 0	76. 4	73. 4	74. 1	76. 3	79. 1
105. 4	105. 0	104. 4	103. 9	103. 3	102. 8	102. 4
8. 5	8. 4	8. 3	8. 3	8. 2	8. 1	8. 1
38. 1	38. 0	38. 0	38. 0	38. 0	38. 0	38. 0
38. 0	38. 0	38. 0	38. 0	38. 0	38. 0	38. 0
323. 5	323. 6	323. 8	323. 9	324. 1	324. 4	324. 6
317. 2	317. 2	317. 6	317. 6	318. 0	317. 9	318. 4
696. 3	697. 0	698. 1	699. 0	698. 5	697. 9	697. 7
715. 0	715. 0	715. 0	715. 0	715. 0	715. 0	715. 0
74. 9	75. 6	76. 9	78. 6	80. 9	82. 9	84. 2
101. 8	103. 2	104. 1	104. 6	105. 3	106. 3	107. 1
8. 3	8. 4	8. 5	8. 6	8. 7	8. 8	9. 0
39. 0	38. 8	38. 8	39. 1	39. 1	38. 9	39. 0
39. 0	39. 0	39. 0	39. 0	39. 0	39. 0	39. 0
302. 2	303. 0	303. 7	304. 3	304. 9	305. 6	306. 5
315. 4	315. 4	315. 0	315. 0	314. 5	314. 5	314. 1

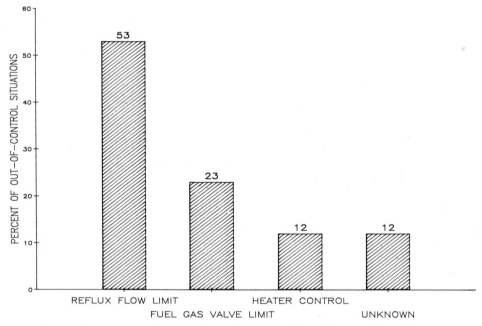

Figure 9. Pareto Chart -- Out Of Control Conditions Unit 10

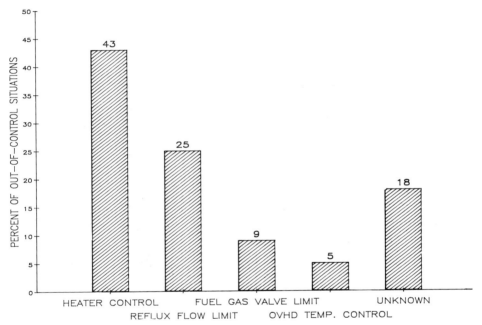

Figure 10. Pareto Chart-- Out Of Control Conditions Unit 11

As indicated by the Pareto for Unit 10, 53% of the out of control situations are a result of heat removal limitations at high light product feed rates. Additional heat exchange would resolve this problem. The Pareto for Unit 11 indicates that 43% of the out of control situations are due to heater control problems. Improvements of this control system would reduce this occurrence. An additional 25% of the out of control situations are due to heat removal limitations at light product feed rates. Again, additional heat exchange would resolve this problem.

Although it did not occur over the duration of this study, another significant out of control situation this statistical control scheme is designed to detect is gradual degradation of controller performance. This may be indicated by the mean value of the control errors gradually moving away from zero.

Maintaining the CUSUM on the natural log of the variances could also help detect loss of controller performance.

Future   Work

One solution to almost all of the out of control problems is to reduce the unit feed rates. Using the data collected, a cost analysis could be made to determine whether lowering feed rates or adding additional equipment is most economical.

It is also important to note that the entire control scheme is based on a calculated cutpoint. Given a repeatable, reproducible, stable gage, laboratory defined cutpoints should be compared to the calculated values and the difference should be control charted. Doing this, the cutpoint calculation could be monitored for accuracy. Further, out of control conditions which develop could be used to highlight potential problems with the temperature and pressure indicators and/or the laboratory gage.

As an additional option, an expert system could be added to analyze out of control conditions and make recommendations to operators for corrective action.

In light of this successful application, other advanced control loops should be added to the monitoring system.

Summary

This Shewhart-CUSUM statistical monitoring technique detects out of control situations well for this advanced control function. The results can be used to prioritize/focus process improvement efforts and develop operation reaction plans for the operators. The results can also be used to define controllers in need of improvement. We believe the wide spread use of this statistical technique could substantially improve the operation, profitability, and safety of continuous processes.

References

Box, G.E.P. and Jenkins, G.M., (1976) Time Series Analysis: Forecasting and Control, (Revised Edition), Holden Day, San Francisco, pgs. 287-293.

Lucas, J.M., (1982), "Combined Shewhart-CUSUM Quality Control Schemes", Journal of Quality Technology, Vol.14, No. 2 pgs. 51-59.

Montgomery, D.C. and Friedman, D,J. (1988) "Statistical Process Control in a Computer Integrated Manufacturing Environment", Statistical Process Control in Automated Manufacturing, J.B. Keats and N.B. Hubele (Eds.), Marcel Dekker,New York,67-87.

Perry, R.H. and Chilton, C.H., (1973) Chemical Engineers' Handbook, Fifth Ed., McGraw-Hill, New York, p.22-17.

SAS Statistical Software, SAS, Inc. Cary, NC.

6
REAL-TIME PROCESS QUALITY CONTROL IN A COMPUTER INTEGRATED MANUFACTURING ENVIRONMENT

Steven A. Yourstone
Anderson Schools of Management
University of New Mexico
Albuquerque, NM   87131

Abstract

This paper describes the development, testing, and analysis of a real-time process quality control algorithm. A time series approach is utilized to model and monitor all of the real-time process quality data for a discrete manufacturing process. A time series approach is implemented since the data stream is likely to be serially correlated. The serial correlation in the data is utilized to identify, estimate, and implement an ARIMA model. The implementation algorithm is suitable for use in a computer integrated manufacturing environment for the purpose of detecting and signaling the presence of assignable causes that result in shifts in the mean, variance, or the autocorrelative structure of the real-time quality data.

Introduction

What makes the present approach unique is that all of the real-time data is analyzed by the algorithm. This means that quality data has been collected on every unit produced in time order of production. This approach contrasts with the methodology of periodic sampling and rational subgrouping of

the quality data. Traditional techniques typically assume that
the samples are uncorrelated. If the samples are correlated
traditional techniques break down. In addition, traditional
techniques will not detect a model shift. A model shift would
be, for example, a shift from an ARMA (2,0,1) model to an AR(2)
model. A pure model shift has taken place when the correlative
structure of the process quality data has shifted -- without a
shift in the mean. Such a model shift would exhibit itself as the
development of significant new structure in the residuals. This
structure shows up most dramatically in the real-time sample
autocorrelation function and in the Portmanteau test.

Why Use All of the Real-Time Data? A sensor can frequently
supply a process quality control tool with one hundred percent
of the real-time data. The question is: What do we gain by
using all of this information? If we consider the procedure of
periodic sampling of process quality data and subsequent
analysis of that data, three problems come to mind
immediately. First, there has been a time lag between the
production of the item and the analysis of the data. Second,
during the interval between samples a temporary shift in the
process may have taken place. Finally, having one hundred
percent of the real-time data means that information on trends,
cycles, and seasonality is not lost due to a choice of some
sampling interval which can mask such phenomena. It is
demonstrated in this paper that to not utilize one hundred
percent of the real-time observations is to throw away
potentially valuable information on the process.

The Time Series Approach

The time series approach to quality control does not assume
independence between or within samples. The dependency in
the data is modeled with a Box-Jenkins ARIMA model (See Box
and Jenkins [1976]). The ARIMA model employed by the
algorithm is based on all of the univariate quality data
generated by a process over a period of time. This means that
no observations are omitted due to the use of periodic sampling.
It is necessary to utilize every observation in order to capture
the correlative structure in the data. ARIMA models estimate
the autocorrelative structure of the underlying stochastic
process. The autocorrelation function is a key tool in the
approach to ARIMA model building.

<u>The Autocorrelation Function</u>   In order to identify, estimate, and utilize an ARIMA model the autocorrelation coefficients must be calculated.   If we had knowledge of the true values of the parameters of the underlying stochastic process we could then calculate the true autocorrelation coefficient at lag k as

$$r = E[(Z_t - m)(Z_{t-k} - m)]/\{E[(Z_t - m)^2(Z_{t-k} - m)]^2\}^{1/2} \qquad (1)$$

where Z_t is the observation at time t, and m is the mean of the process.   Since in practice we typically do not know the true mean of the process we then must use an estimated (or sample) autocorrelation function (SACF) to identify the correct ARIMA model to use.   The estimated autocorrelation of the data at lag k, $\hat{r}_k$, is

$$\hat{r}_k = C_k/C_0 \qquad (2)$$

where k = 1,2, 3... K.

$$C_k = (1/N)\sum_{t=1}^{N-k} (Z_t - \overline{Z})(Z_{t-k} - \overline{Z})$$

where N is the number of observations used in fitting the ARIMA model, k is the lag of interest, K is the total number of lags used in the autocorrelation function, and $\overline{Z}$ is the sample average of the observations.

<u>The Estimated Model of the Process</u>   Equation (2) is used in combination with an estimate of the partial autocorrelation function (See Box and Jenkins [1976] for details) to identify a candidate ARIMA model.   The parameters of the candidate ARIMA model can be estimated using a non-linear least squares algorithm.   If the candidate model passes the diagnostic testing stage then the model can be utilized for real-time control.   We may forecast the next value for the process quality at time t with an ARIMA model which may take the form

$$Z_t = \hat{\mu} + \hat{\beta}_1 Z_{t-1} + ... + \hat{\beta}_p Z_{t-p} + \Delta$$
$$-\hat{\theta}_1 a_{t-1} - ... - \hat{\theta}_q a_{t-q} \qquad (3)$$

where $\hat{\mu}$ is the estimate of the mean of the underlying process,

$\hat{\beta}_p$ is the estimated p-th order autoregressive coefficient, the symbol delta, Δ, represents a trend component, $\hat{\theta}_q$ is the estimated q-th order moving average coefficient, and a_t is the residual at time t.   One can easily see that the forecast of the next observation, Z_t, is a combination of the estimated mean, plus a weighted combination of prior observations on the data stream, plus some constant, plus some weighted combination of the prior residuals.

In addition to modeling the mean and the variance, the ARIMA time series model also estimates the autocorrelative structure of the process quality data.   Traditional Shewhart types of control charts monitor the mean or the variance of the process quality data.   Shewhart charts do not model the correlative structure of the data.   The estimated autocorrelative structure of the data provides additional information on the process quality.

Forecasting equation (3) is utilized in the REAL-TIME SPC algorithm to forecast the next observation on the process quality data.   The difference between the predicted value for the next observation and the actual value of the next observation is the residual for the model.   The algorithm utilizes the stream of residuals over time to screen the process for indications of assignable causes of variation.

The Residual as the Control Variable

The REAL-TIME SPC algorithm works with the estimated autocorrelation function of the real-time residuals.   A window of fifty residuals at a time forms the basis for estimating the current sample autocorrelation at each lag of interest.   The sample autocorrelation function (SACF) of the residuals as derived from Equation (2) is

$$\hat{r}_k = \sum_{i=1+k}^{50} (a_i - \bar{a})(a_{i-k} - \bar{a}) / \sum_{i=1+k}^{50} (a_{i-k} - \bar{a})^2 \tag{4}$$

where $\hat{r}_k$ is the estimated k-th lag autocorrelation coefficient of the residuals, a_i is the i-th residual, $\bar{a}$ is the actual mean of the real-time residuals estimated in the first window of fifty, and the summation is over the window of fifty residuals. If the ARIMA model we fit to the process, quality data were the true model of the underlying stochastic process then the residuals would follow a normal distribution with zero mean and constant variance (Anderson [1942]). Theoretically then, all lags of the residuals should then display autocorrelations equal to zero if chosen ARIMA model is an optimal fit. If the process quality data does not undergo any shifts in the mean, the variance, or the autocorrelative structure while we are monitoring it, then the residuals will exhibit a mean of approximately zero and nonsignificant autocorrelations at all lags.

An Exponentially Smoothed Estimate of the SACF   The REAL-TIME SPC algorithm plots an exponentially smoothed estimate of a specific lag k sample autocorrelation of the real-time residuals. The estimate of the sample autocorrelation, $\hat{r}_k$, at a given lag k is composed of a smoothing constant times the current estimate of the autocorrelation at lag k, plus one minus the smoothing constant times the estimate of the lag k autocorrelation computed in the previous window of fifty residuals. The statistic plotted on the chart is

$$S_w(\hat{r}_k) = \alpha\hat{r}_k + (1-\alpha) S_{w-1}(\hat{r}_k) \qquad (5)$$

where $\hat{r}_k$ is the current estimate of the lag k autocorrelation and a is the smoothing constant, which lies in the interval $0<a<=1$. The w subscript denotes the current window of fifty residuals. The residuals should have autocorrelations at all lags as close to zero as possible if the process is operating in a state of statistical control. The lag k residual autocorrelation will fluctuate from window to window of observations. This fluctuation is bounded by an upper and lower control limit calculated in basically the same way as the limits for a geometric moving average (GMA) control chart. See Roberts [1959] and Montgomery [1985] for a discussion of the geometric moving average control limit calculation. The formula for calculating the control limits for the exponentially estimated autocorrelation function (ACF) is

$$0 \pm D[(1/(50-k)]^{1/2}[(a/(2-a)]^{1/2} \qquad (6)$$

where k = the lag, a = the smoothing constant input by the user, and D = the multiple of the standard deviation used to set the control limit width. The square root of the quantity 1/(50-k) is the estimate of the standard deviation of the lag k sample residual autocorrelation.

Information Provided by the SACF of the Residuals The sample autocorrelation function (SACF) of the real-time residuals is a key module of the REAL-TIME SPC algorithm. This module allows the algorithm to capture information on the process quality data that a traditional approach consisting of taking samples at some interval will not. The additional information provided is the correlation between the quality dimension of a part that was just manufactured and one that there may be assignable causes of variation operating in the process that may require investigation. A shift in the autocorrelative structure of the underlying data stream will manifest itself as new autocorrelative structure in the real-time residuals.

An Overview of the Algorithm

Figure 1 provides an overview of the way in which the REAL-TIME SPC algorithm is applied to a process. The procedure begins with a process sensor. The process sensor captures all of the univariate observations on the quality dimension of a discrete part. When the process is known to be operating in a state of statistical control, an ARIMA model can be fit to a sample of at least fifty observations.

The ARIMA model specification is then input to the REAL-TIME SPOC algorithm for use in real-time monitoring of the process. The algorithm makes a forecast for the next observation on the quality data. When the sensor obtains the next actual observation, a residual is then calculated. This residual is then included in a moving window of fifty estimated residuals. This window of fifty residuals forms the input stream for a series of control modules employed by the algorithm. Several of these modules plot some function of the real-time residuals.

The ARIMA model characterizes the process in its in-control state. When assignable causes develop, they will be manifested as a shift in the data stream from the original ARIMA model

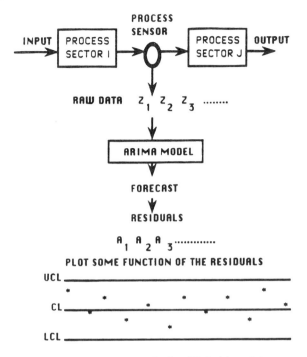

Figure 1. Overview of the SPC Algorithm

specification. This shift is directly translated into significant autocorrelative structure in the real-time residuals. The algorithm is designed to detect shifts in the stream of real-time residuals.

Detection of Mean and Autocorrelative Shifts   The sample autocorrelation control chart detects shifts in the mean as well as shifts in the autocorrelative structure. This result can be seen by examining the equation for the value of the mean of a simple autoregressive process of order two. The term 'order two' corresponds, in this case, to a time series model with two autoregressive terms. The mean of such a process is (For details see Montgomery and Johnson [1976])

$$m = x/(1-b_1-b_2) \qquad (7)$$

where x is some constant, and the b coefficients are the first and second-order autoregressive coefficients. It is clear that a shift in the mean will translate itself into a shift in the coefficients of the data stream in which the shift takes place, and hence in the dynamic behavior of the window of fifty real-time residuals. This shift in the coefficients represents a shift in the correlative structure of the residuals. When a pure shift in the autocorrelative structure takes place, there is no shift in the mean. It is in such a situation that the REAL-TIME SPC algorithm excels and traditional approaches do not perform well.

Control Modules   The algorithm works with a moving window of fifty residuals for as long as the process continues to generate observations. The REAL-TIME SPC algorithm implements several control modules. The control modules include: the sample autocorrelation function, the Portmanteau test, the geometric moving range, and the geometric moving average control modules for the purpose of monitoring the residuals for dynamic shifts in the process. With the exception of the Portmanteau test, each of these control modules utilizes the concept of exponential weighting. The current estimate of the statistic plotted on the chart is composed of some weight times a current value plus one minus that same weight times the previous estimate of the exponentially smoothed statistic.

The Portmanteau Test   The Portmanteau test provides additional information on the correlative structure of the process quality data. It is a complementary test to that provided by the sample autocorrelation function. The Portmanteau test works with the sample autocorrelations over all lags, k. Departures of the process quality data from the original model will strongly affect the Portmanteau test. A shift in the mean, the variance, or the autocorrelative structure of the process quality data will be detected by this test. The Portmanteau test implements a statistic developed by Ljung and Box [1979] for a set of m sample autocorrelations. The Ljung and Box statistic is

$$S(\hat{r}) = n^2 \sum_{k=0}^{m} (n\text{-}k)^{-1}(\hat{r}_k)^2 \tag{8}$$

where n is the number of observations (here equal to fifty residuals) and m is set equal to twelve (for the number of lags of sample autocorrelations used). The value calculated using (8) is then compared with the chi-square value found in a standard chi-square table (refer to Box and Jenkins [1976]).

Geometric Moving Range Control Chart   The geometric moving range between successive pairs of residuals is used to track the dispersion of the process quality data in the REAL-TIME SPC algorithm.   The geometric moving range allows the user of the algorithm to alter the sensitivity of the moving range filter through adjustments to the smoothing constant.   The rationale behind the use of the geometric moving range is to increase the ability of the moving range to detect small shifts in the process variability.   For the case of small samples, a simple range chart is relatively insensitive to small shifts in the process standard deviation.   Duncan [1951] provides the operating characteristic curves for range charts.   In short, these operating characteristic curves for the range chart say that the probability that the chart will detect a shift in the standard deviation of a process decreases as we look for smaller shifts.   The REAL-TIME SPC algorithm uses the concept of exponential weighting to improve the performance of the moving range in detecting small shifts in the process standard deviation since samples of size one, and moving ranges of only two successive pairs of residuals, enter in to the calculation.   The statistic plotted on the geometric moving range control chart is

$$GMR_t = aR_t + (1-a)GMR_{t-1} \qquad (9)$$

where GMR$_t$ is the geometrically smoothed estimate of the moving range of successive pairs of time-ordered residuals, the constant, a, lies in the interval $0 < a <= 1$, and R$_t$ is the range for the most recent time-ordered pair of residuals at time t.   The REAL-TIME SPC algorithm works with moving ranges of successive pairs of residuals for every window of fifty residuals. The control limits for the moving range control chart are given by (10).

$$UCL = \bar{R} + D\hat{\sigma}_{mr}[(\alpha/(2-\alpha))]^{1/2}$$
$$LCL = \bar{R} - D\hat{\sigma}_{mr}[(\alpha/(2-\alpha))]^{1/2} \qquad (10)$$

The center line for the control chart, $\bar{R}$, is equal to the average of the moving ranges, estimated in the first window of fifty residuals. D represents the width of the control limits, in units of standard deviation of the statistic we plot on the chart. $\hat{\sigma}_{mrr}$ is an estimate of the standard deviation of the moving ranges of the residuals, estimated of the standard deviation of the geometric moving range is obtained from $\hat{\sigma}_{mrr} = [d_3\overline{R}/d_2]^{1/2}$. The values for d_3 and d_2 may be found in most standard quality control texts (see, for example, Montgomery [1985]).

Geometric Moving Average Control Chart   The geometric moving average control chart (GMA) is utilized for the purpose of detecting shifts in the mean of the process quality data.   Roberts [1959] demonstrated that the GMA is a weighted average of all of the prior observations (sample means if using rational subgrouping, with sample sizes greater than one) on a process. Roberts [1959] demonstrated that the GMA chart is well-suited for the purpose of early detection of small shifts -- simply by setting the smoothing constant to smaller values.   The sensitivity of the chart is linked to the "memory" of the chart and refers to the ability of the chart to be more or less influenced by prior observations through adjustments to the smoothing constant.   The statistic plotted on the GMA control chart, when it is used to   monitor sample averages, is

$$Z_t = \alpha\overline{X}_t + (1-\alpha)Z_{t-1} \tag{11}$$

where the smoothing constant, a, lies in the interval $0<a,=1$. Equation (11) is modified for use in tracking the residuals of the ARIMA model for the process quality data.   In order to utilize (11) for the purpose of tracking the average value of the real-time residuals, we simply replace $\overline{X}_t$ with a_t, the residual at time t.   The control limits for the geometric moving average control chart are

$$UCL = \bar{a} + D(R/d_2)[(a/(s-a)]^{1/2}$$
$$LCL = \bar{a} - D(R/d_2)[(a/(s-a)]^{1/2} \tag{12}$$

The center line for the chart, $\bar{a}$ is the estimated average of the real-time residuals for the first window of fifty.   The TIME SPC algorithm.   The response variable in all of the experiments was the run length behavior for the REAL-TIME

SPC algorithm. Furthermore, the results obtained from the experiments performed on the algorithm were compared with those obtained in three experiments performed on the Shewhart $\bar{X}$ chart.

The first experiment was designed to investigate the effects of two factors, at three levels, on the in-control run length of the REAL-TIME SPC algorithm. A fixed effects factorial design with two factors each at three levels was implemented. The first factor was the level of the smoothing constants in the geometric moving average (GMA) and geometric moving range (GMR) control modules. The second factor was the width of the control limits for the geometric moving average and the geometric moving range control modules.

The second experiment was a fixed effects factorial design which investigated the effects of three factors, at three levels, on the out-of-control run length of the algorithm. Two of the factors were the same as in the first experiment. In addition, shifts in the mean were included as a third factor in the second experiment.

The third experiment was a fixed effects factorial design with two factors at three levels. This experiment was run for the purpose of determining the in-control run length for a Shewhart $\bar{X}$ chart on the same data set used to test the algorithm. The two factors were the width of the control limits and the subgroup size.

The fourth experiment was run on the Shewhart $\bar{X}$ chart. This experiment was performed for the purpose of determining the out-of-control run length behavior of the $\bar{X}$ chart when faced with the same data set as the REAL-TIME SPC algorithm. This experiment was a fixed effects factorial design with three factors each at three levels. The three factors were the width of the control limits, the subgroup size, and the size of the shift in the mean.

The fifth experiment was run to determine the run length behavior of the algorithm in the presence of shifts in the form of the model. A shift from an ARIMA (2,0,1) process to an AR(2) was examined. This shift involves a change in the autocorrelative structure of the process quality data--without a change in the mean of the process. A fixed effects factorial design with two factors at three levels was examined. The first factor was the width of the control limits for the sample autocorrelation function module, the geometric moving average module, and the geometric moving range module. The second factor was the value for the smoothing constants in each of these modules. The experiment began with the process being well modeled by an AR(2,0,1) with a subsequent process shift to an AR(2) at observation number 51.

The sixth experiment was run to determine the run length behavior of the Shewhart X chart in the face of shifts in the correlative structure. In such a shift we assume that the mean of the process has remained the same. The shift in the correlative structure is identical to that in design number five. The sixth experimental design was a fixed-effects factorial design with two factors at three levels. The factors were the width of the control limits and the subgroup size.

All experimental designs were run with the first 50 observations from the process known to be drawn from the process when it was in a state of statistical control. At precisely observation number fifty-one the shift was introduced to the control scheme. Both the algorithm and the Shewhart X chart were found to compete on approximately the same terms. Both were faced with the same data set and process upsets. However, the Shewhart chart was given an advantage in terms of earlier detection, in that it was not implemented with periodic sampling. That is, the Shewhart chart was not required to wait for a specified interval of observations before drawing another rational subgroup of data.

Interpretation of Results These experiments were run with a single replication. It is necessary to run more than one replicate in an experimental design in order to estimate the mean square (or experimental) error. The mean square error is utilized for performing F-tests of the significance of main effects and interaction effects in a designed experiment. However, in the present experiments, replication was not necessary since there was no experimental error. The data set is known, and the point at which shifts occur is also known. If a test combination on the algorithm were run a second time identical results would be obtained.

The method of analysis then becomes a simple comparison of the total sums of squares of each factor. We can then obtain a measure of the relative importance of each factor. One might argue that this does not constitute a true test of statistical significance. However, it does convey a valid measure of the relative contribution of each factor to the run length behavior of the algorithm and the Shewhart $\bar{X}$ chart, and it does convey to the practitioner what must be done to adjust the sensitivity of the algorithm to a specific application.

There were a total of ninety simulation runs over six different experiments. The results are summarized in the next section.

Results of the Designed Experiments

Design 1: In-Control Run Length for the Algorithm The value of the smoothing constants in the geometric moving range (GMR) and the geometric moving average (GMA) control modules had the greatest impact in determining the run length behavior of the algorithm when the process was known to be operating in a state of statistical control. Or as some authors would prefer to state it, the process was operating without assignable causes of variation.

Design 2: Out-of-Control Run Length for Mean Shifts for the Algorithm Shifts in the mean also translate into shifts in the autocorrelative structure of the data. The algorithm detects shifts in the mean with the geometric moving average control scheme, the sample autocorrelation control scheme, and the Portmanteau test. An out-of-control signal from both the geometric moving average control scheme as well as the sample autocorrelation control scheme is a strong indication that there has been a shift in the mean of the process. Shifts in the mean of the process (non-stationary mean) are also indicated when the Portmanteau test is significant. A signal from only the sample autocorrelation function module is an indication that the autocorrelative structure has shifted--but not the mean. Table 1 lists the number and percentage of simulation runs where

Table 1. Mean Shift Detection By the Algorithm

Filter Detecting Shift	Number of Times Filter Detected First	Percent of Total	Size of Shift
Autocorrelation Filter	10 of 27 runs	37.04%	4 at 1.5σ 6 at 0.5σ
Portmanteau & Autocorrelation Filter	3 of 27 runs	11.11%	3 at 2.5σ
Geometric Moving Average Filter	5 of 27 runs	18.52%	2 at 1.5σ 3 at 2.5σ

each of the filters in the algorithm detected the shift in the mean first.   Note that in 37.04% of the simulation runs the autocorrelation filter detected the shift first.

A total of 18 out of 27 or 66.67% of the simulation runs resulted in correct and early detection of a mean shift.   In the remaining 9 runs, premature out-of-control signals were given due to the use of very tight control limits and/or smoothing constants that were too low.   A smoothing constant of 0.3 was responsible for a false signal rate of 88.89% of the total simulation runs in which it was used.

For a smoothing constant of 0.6, a false signal rate of 33.33% was realized in the simulation runs in which it was utilized. Whereas, for a smoothing constant of 0.9 none of the simulation runs resulted in a false signal.Figure 2 illustrates the appearance of the geometric moving average of the algorithm for a shift of 0.5 standard deviations in the mean.   The shift in the mean occurs at observation number 51.   The column of the output labeled "CUMULATIVE OBSERVATION NUMBER" gives the number of the observation since production began for that manufacturing run.   The column labeled "OBSERVATION NUMBER" gives the position of that observation in the moving window of 50 observations.   It should be noted that these

OBSERVATION NUMBER	GMA VALUE	CUMULATIVE OBS.NO	LCL CL UCL
1.00	-.0071	40)	: * . :
2.00	-.0099	41)	: * .
3.00	.0505	42)	: . : *
4.00	-.0013	43)	: *.
5.00	.0021	44)	: *
6.00	.0086	45)	: . *
7.00	.0162	46)	: . *
8.00	.0092	47)	: . *
9.00	.0369	48)	: . * :
10.00	.0273	49)	: . *
11.00	.0038	50)	: .*
12.00	.0347	51)	: . * :
13.00	.0368	52)	: . * :
14.00	.0426	53)	: . *:
15.00	.0297	54)	: . * :
16.00	.0449	55)	: . *
17.00	.0511	56)	: . : *
18.00	.0280	57)	: . * :
19.00	.0288	58)	: . * :
20.00	.0394	59)	: . * :
21.00	.0488	60)	: . :*
22.00	.0221	61)	: . * :
23.00	.0264	62)	: . * :
24.00	.0181	63)	: . * :
25.00	.0099	64)	: . * :
26.00	.0268	65)	: . * :
27.00	.0133	66)	: . * :
28.00	.0178	67)	: . * :
29.00	.0148	68)	: . * :
30.00	.0391	69)	: . * :
31.00	.0385	70)	: . * :
32.00	.0130	71)	: . * :
33.00	.0078	72)	: . * :
34.00	.0009	73)	: * :
35.00	-.0019	74)	: *.
36.00	-.0079	75)	: *
37.00	.0044	76)	: .*
38.00	.0045	77)	: .*
39.00	.0247	78)	: . *
40.00	-.0069	79)	: * .
41.00	-.0051	80)	: * .
42.00	-.0330	81)	: * .
43.00	-.0104	82)	: * .
44.00	.0036	83)	: .*
45.00	.0042	84)	: .*
46.00	.0143	85)	: . *
47.00	-.0062	86)	: * .
48.00	-.0193	87)	: * .
49.00	-.0048	88)	: * .

Figure 2. Example output for GMA control module

observations are not the raw data, but rather they are the residuals.

Design 3: In-Control Run Length for the Shewhart X̄ Chart   Nine simulation runs were performed on the Shewhart X̄ chart to analyze the in-control behavior.   In three out of the nine simulation runs premature false out-of-control signals were generated.

Design 4: Out-of-Control Run Length for the Shewhart X̄ Chart for Shifts in the Mean.   The results of the twenty-seven simulation runs for a mean shift for the X̄ chart are very interesting.   Eleven, or 40.74%, of the simulation runs resulted in false   out-of-control signals before the shift at observation number fifty one was ever encountered.   In five, or 18.5%, of the simulation runs the shift in the mean was never detected.

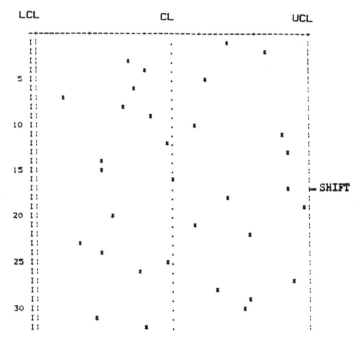

Figure 3. X̄ control chart for a shift in the mean of 0.5 Standard Deviation

A shift of 0.5s, with control limits set at 3.0s and a subgroup size of three, went undetected. Figure 3 illustrates the appearance of the X̄ chart when applied to real-time data. The numbers on the left side of the plot indicate the sample number. The shift in the mean of 0.5 standard deviations does not appear on the chart. The advantage of using a geometric moving average to detect small shifts in the mean is clear.

<u>Design 5:   Out-of-Control Run Length for Shifts in the</u> <u>Autocorrelative Structure for the Algorithm</u> The use of a smoothing constant lower than 0.9 in the sample autocorrelation function module leads to a high false alarm rate. In five out of the six runs having a smoothing constant of 0.8, false alarms occurred in the first window of fifty observations in the sample autocorrelation function module--before the shift had occurred.

The optimal parameter settings to use, in the case of the piston diameter data, were found to be limits of 3.5s for the sample autocorrelation function module and smoothing of 0.9. The shift in the autocorrelative structure was detected within three observations following the shift. It should be kept in mind that minimizing the out-of-control run length is only one half of the decision in setting the correct parameters. The other half of the decision is to achieve an acceptable level for the in-control run length of the control scheme. A setting of 3.0s for the sample autocorrelation function module limits, coupled with a smoothing constant of 1.0, will provide a longer in-control run length, while still detecting the shift within five observations following the shift--in the case of the piston diameter data.

<u>Design 6:   Out-of-Control Run Length for Shifts in the</u> <u>Autocorrelative Structure for the Shewhart X̄ Chart</u>   The performance of the Shewhart X̄ chart was quite poor when faced with a shift in the autocorrelative structure at observation number fifty one. Out of a total of nine simulation runs on the X̄ chart, 3 or 33.3%, led to false out-of-control signals before the shift ever occurred. In the remaining 6 runs, or 66.7% of the total runs, the shift was never detected. In one hundred percent of these nine simulation runs there were either false alarms or the shift was missed entirely. Figure 4 illustrates the appearance of the X̄ chart for a shift in the autocorrelative structure of the process quality data. The numbers listed on the

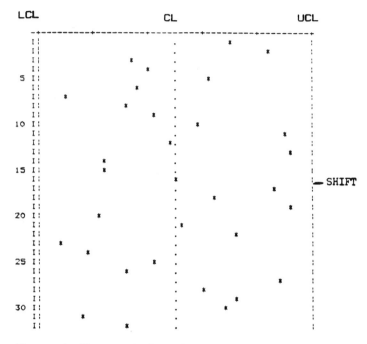

Figure 4. X̄ control chart for a correlative shift.

left side of the plot indicate the sample number.   No indication of a shift is apparent from an inspection of the plot.

<u>Summary of the Experiments</u>   It is clear that the Shewhart X̄ chart did not perform well on the original data stream when autocorrelation was present.   Table 2 compares the results of experiments involving shifts in the mean.   The Shewhart X̄ chart suffers from a higher false alarm rate for small and large shifts in the mean.

It should be noted that the false alarm rates for both the algorithm and the Shewhart X̄ chart are somewhat inflated due to the nature of the factorial designed experiments.   The fixed levels of each of the factors used in the experimental designs were selected so that a sufficient range of sensitivity of the two quality control techniques could be elicited.   For a given data set, the algorithm and the Shewhart X̄ chart could be designed

Table 2

SUMMARY OF RESULTS FOR SHIFTS IN THE MEAN: THE
OBSERVATION AT WHICH THE SHIFT WAS DETECTED
AND
THE FALSE SIGNAL FREQUENCY

SIZE OF SHIFT	ALGORITHM DETECTED SHIFT	ALGORITHM FALSE ALARM RATE	X DETECTED SHIFT	FALSE ALARM RATE
0.5 s	n=59	33.33%	n=66	77.78%
1.5 s	n=54	44.44%	n=55	44.44%
2.5 s	n=54	33.33%	n=55	55.56%

to elicit a higher in-control run length as well as a lower out-of-control run length.   What is clear from this research is that the Shewhart $\bar{X}$ chart, or any technique that assumes uncorrelated inputs, should not be used to monitor real-time process quality data when one hundred percent sampling is being used.

Conclusions

This research demonstrates that the time series approach to real-time process quality control should be used when working with one hundred percent of the quality data captured form a process.   Traditional approaches break down in the face of one hundred percent of the data, since they assume independence both between and within the samples.   The algorithm reported in this research is a viable basis for establishing a statistical process control system in a sensor-driven environment.

References

Anderson, R.L. (1942)   Distribution of the Serial Correlation Coefficient, Annals of Mathematical Statistics. 13. 1.

Box, G.E.P., and David A. Pierce (1970), "Distribution of Residual Autocorrelations in Autoregressive-Integrated Moving

Average Time Series Models", Journal of the American Statistical Association. Vol. 65. No. 332. December: 1509-1526.

Box, G.E.P., and G.M. Jenkins. (1976) Time Series Analysis: Forecasting and Control. Oakland, California: Holden Day.

Duncan, Acheson J. (1974) Quality Control and Industrial Statistics. Homewood, Illinois: Irwin.

Duncan, Acheson J., (1951). Operating characteristics of R Charts, Industrial Quality Control. March: 40-41.

Ljung, G.M., G.E.P. Box. (1979). On a Measure of Lack of Fit of Time Series Models Biometrika. Vol. 65. 2: 297-303.

Montgomery, Douglas C. (1985) Introduction to Statistical Quality Control,. New York: Wiley.

Montgomery, Douglas C. (1984). Design and Analysis of Experiments,. New York: Wiley.

Montgomery, Douglas C. and Lynwood A. Jonson (1976) Forecasting and Time Series Analysis, New York: McGraw-Hill.

Oakland, Jonn S. (1986). Statistical Process Control London: Heinemann.

Roberts, S.W. (1959) "Control Chart Tests Based on Geometric Moving Averages." Technometrics. Vol. 1. No. 3. August: 239-251.

7
ADAPTIVE CONTROL CHARTS

John J. Flaig*
Apple Computer, Inc.
10201 North De Anza Blvd., MS 23AP
Cupertino, CA 95014

Introduction

Classical control chart methods are sub-optimal with respect to two very important criteria -- sampling and use of data. Specifically, the chart's decision criteria does not use all the data available and the sampling rules remain fixed no matter what data trends may develop. For example:

The decision rules use only a limited history (i.e., nine points above the mean) and the sampling procedures are fixed (i.e., subgroup size is constant).

Cusum and EWMA charts are designed to improve chart performance relative to the complete use of data. But, in general, we have not seen self-modifying control chart schemes designed to address the issue of responding to data trends. The goal of this paper is to propose one such method.

The "Adaptation Principle" is a simple methodology designed to link the sample size to whatever data trends might develop and which can be applied to various control charting schemes. The effect of this modification being to substantially increase the sensitivity of the chart with only a moderate increase in inspection.

*Current affiliation: Apple Computer, Inc., 3585 Monroe St., Santa Clara, CA 95051

111

The Adaptation Principle

To illustrate one of the several possible ways the adaptation principle might be applied to a mean control chart, assume that we have verified the four control chart requirements:

The subgroup size is rational.

The measurement tool is capable.

There is independence both within and between subgroups.

The distribution of sample statistics is normal.

Further, let the subgroup size be n and the distribution of the sample means be normal with mean m and standard deviation s_m (i.e., $N(m, s_m{}^2)$) where $s_m = s/\sqrt{n}$).

Next set the control chart standards m" and s" to   m" = m and s"=$s_m \sqrt{n}$. The chart is then established and we begin to plot new points. If a point is within the mean ± 1 sigma, then the sample size of the next subgroup is n. If a point is outside the mean ± 1 sigma, but inside the mean ± 2 sigma, then the sample size of the next subgroup is increased to 2n. If the point is outside the mean ± 2 sigma, but inside the mean ± 3 sigma, then the sample size of the next subgroup is increased to 3n.

In other words, using standard control chart terminology, if the point is in zone C, then the sample size for the next subgroup is n. For a point in zone B, the next sample size is 2n, and for a point in zone A, the next sample size is 3n. We shall designate such plans as AM(1n,2n,3n). This notation can be generalized so that plans which divide the normal distribution into three zones are denoted by $AM(n_1, n_2, n_3)$. Of course different sample sizes (n_i's) and different sample size shift points (z_i's) are possible, the 1-2-3 scheme is used, as an example here, because it is simple and easy to understand.

One of the drawbacks of adaptive control charts is that as the sample size changes so do the upper and lower control limits (UCL and LCL). This creates computational problems as well as visual problems associated with out-of-control pattern recognition (see Figure 1). Fortunately, the computational problems can be overcome by using a programmable calculator or personal computer. The shifting control limits issue is completely removed by plotting standardized values. That is, instead of plotting m_i, plot z_i ($z_i = (m_i-m'')/(s''/\sqrt{n_i})$), where m'' and s'' denote the standard values used to define the control chart).

The adaptation principle can also be applied to Cusum charts by holding the angle q and the subgroup size n fixed, then shortening the lead distance of the V-mask. For example, given d, we create two new V-masks by defining d_1 = .8d and d_2 = .6d. We can now establish a nested set of V-masks such that the least sensitive region is called zone C, the next level, zone B, and finally, zone A   (in an analogous manner to Shewhart charts). Then apply the switching rules as they were described above, for an adaptive mean control chart to the new adaptive cusum chart (denoted $ACS(n_1,n_2,n_3)$), except we now look at all points, not just the most current (see Figure 2). If a shift is

$n_0 = 4$ $n_1 = 8$ $n_2 = 4$

$$UCL_0 = \bar{\bar{x}} + \frac{3\,\sigma_x}{\sqrt{n}} \qquad UCL_1 = \bar{\bar{x}} + \frac{1}{\sqrt{2}}\left(\frac{3\,\sigma_x}{\sqrt{n}}\right) \qquad UCL_2 = \bar{\bar{x}} + \frac{3\,\sigma_x}{\sqrt{n}}$$

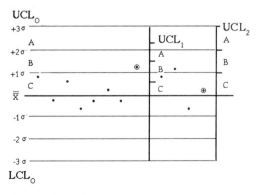

Figure 1. Mean Chart

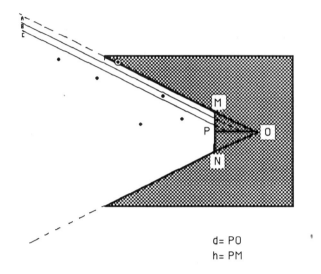

d= PO
h= PM

Figure 2. A V-Shaped Mask for Use on a CUSUM Chart

indicated, say n_1 to n_2, then we compute a new $h_1 = C\,s/\sqrt{n_2}$, where $C = h\,\sqrt{n_1}/s$ The new angle $q_1 = \tan^{-1}(h_1/d)$ will be smaller than the q for n_1.

Performance Measures for Adaptive Control Charts

The speed with which a control procedure detects a process shift is an important measure of its performance. The Average Run Length (ARL) provides such a measure for control charts. The average run length is simply the average number of subgroups one would expect to have to check before detecting a shift of a certain magnitude once the shift has occurred. The detection criteria is assumed to be one point outside the three sigma control limits.

For example, a one sigma shift in mean from 100 to 104.47 in a Shewhart Mean control chart with m"=100, s"=10 and n=8 yields an ARL of 43.94. The same shift to m' = 104.47 using the Adaptive Mean control chart AM(5,10,15) generates an ARL of 28.37. The AM(5,10,15) plan has an Average Sample Number (ASN) of 8.05 which is essentially equal to the mean chart using n=8. Hence, there is about a 35% reduction in ARL, for the same mean shift generated by using the adaptive approach with the same amount of inspection.

A comparative example of Cusum, Shewhart Mean and Adaptive Mean control charts is given Figure 3. and Table 1.

Derivation of ASN and ARL for Adaptive Control Charts

The average sample size (ASN) for an adaptive mean control chart is a function of the mean shift, sample size and the

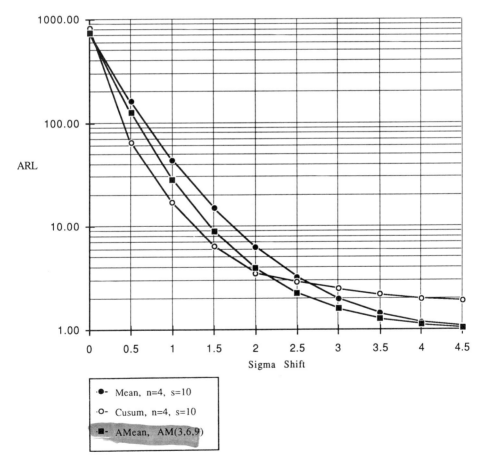

Figure 3. Average Run Length

Table 1   Average Run Length for the Mean Control Chart

	Sample #1	Sample #2	Sample #3
Mean =	100.0000		
Sigma(x) =	10.0000		
n =	4	8	12
Sigma(m) =	5.0000	3.5355	2.8868
UCL =	115.00	110.61	108.66

	Sigma shift m	New mean m'	z	Pa =P(m'<UCL)	ARL =1/(1-Pa)	Cusum n=4	AM(3,6,9) ASN = 4
Sample #1	0	100	-3.00	0.99865	738.06	830.00	740.00
	0.5	102.5	-2.50	0.99378	160.78	65.00	126.23
	1	105	-2.00	0.97724	43.94	17.00	28.37
	1.5	107.5	-1.50	0.93320	14.97	6.40	8.94
	2	110	-1.00	0.84135	6.30	3.50	3.93
	2.5	112.5	-0.50	0.69145	3.24	2.90	2.27
	3	115	0.00	0.50000	2.00	2.50	1.60
	3.5	117.5	0.50	0.30855	1.45	2.20	1.28
	4	120	1.00	0.15865	1.19	2.00	1.13
	4.5	122.5	1.50	0.06680	1.07	1.90	1.05
	5	125	2.00	0.02276	1.02	1.85	1.02
Sample #2		100	-3.00	0.99865	738.06		
		102.5	-2.29	0.98906	91.42		
		105	-1.59	0.94361	17.73		
		107.5	-0.88	0.81021	5.27		
		110	-0.17	0.56812	2.32		
		112.5	0.54	0.29615	1.42		
		115	1.24	0.10699	1.12		
		117.5	1.95	0.02561	1.03		
		120	2.66	0.00395	1.00		
		122.5	3.36	0.00039	1.00		
		125	4.07	0.00002	1.00		
Sample #3		100	-3.00	0.99865	738.06		
		102.5	-2.13	0.98357	60.85		
		105	-1.27	0.89760	9.77		
		107.5	-0.40	0.65612	2.91		
		110	0.46	0.32130	1.47		
		112.5	1.33	0.09173	1.10		
		115	2.20	0.01405	1.01		
		117.5	3.06	0.00110	1.00		
		120	3.93	0.00004	1.00		
		122.5	4.79	0.00000	1.00		
		125	5.66	0.00000	1.00		

Reference: A.J. Duncan, "Quality Control and Industrial Statistics", 5th Ed. page 518.

standards used in the control chart for mean and standard deviation. This relationship may be expressed:

$$ASN = f(m,n: m'',s'')$$

where m is the mean shift measured in sigma-mean units, n is the subgroup size, and m" and s" are control chart standards.

If we partition the normal distribution above the mean into regions called A, B and C, and below the mean into A-, B- and C- respectively, then for an $AM(n_1,n_2,n_3)$ plan we have:

$$ASN(m) = [P(A)+P(A-)] \cdot WA + [P(B)+P(B-)] \cdot WB +$$
$$[P(C)+P(C-)] \cdot WC \qquad (1)$$

where

$$P(A) = \int_{2-m}^{3-m} \emptyset(z)dz \quad P(A-) = \int_{-3-m}^{-2-m} \emptyset(z)dz \quad P(B) = \int_{1-m}^{2-m} \emptyset(z)dz \quad P(B-) = \int_{-2-m}^{-1-m} \emptyset(z)dz \quad ,$$

$$P(C) = \int_{-m}^{1-m} \emptyset(z)dz \quad P(C-) = \int_{-1-m}^{-m} \emptyset(z)dz \quad ,$$

and the Wi's are the weighting factors. For an AM(1n,2n,3n) plan, the Wi's are:

WA = 3n, WB = 2n and WC = 1n.

For example, for m=0, m=1 and m=3:

$$ASN(0) = 3n(.0428) + 2n(.2718) + n(.6826)$$
$$= 1.35n$$

$$ASN(1) = 3n(.1359 + 0) + 2n(.3413 + .0214) + n(.3413 + .1359)$$
$$= 1.61n$$

$$ASN(3) = 3n(.3413 + 0) + 2n(.1359 + 0) + n(.0214 + 0)$$
$$= 1.32n$$

To determine the average run length (ARL) for an adaptive control procedure one must first find the probability of accepting a sample as "in-control", when in fact the mean has shifted by m-sigma. This probability shall be denoted by P_a below.

Assume that the process is producing units and that at some time (t), the mean shifts to a new level m' (m' = m" + m $\cdot (s''/\sqrt{n_1})$). Further, let m_1 be the subgroup mean of the sample just before the shift and let m_2 be the subgroup mean just after the shift, then:

$P_a=$ P(m_1eA) and P(m_2<UCL$_A$)
   or P(m_1eA-) and P(m_2<UCL$_A$)
   or P(m_1eB) and P(m_2<UCL$_B$)
   or P(m_1eB-) and P(m_2<UCL$_B$)
   or P(m_1eC) and P(m_2<UCL$_C$)
   or P(m_1eC-) and P(m_2<UCL$_C$)

Which is:

$P_a=$ [P(A) + P(A-)] $\cdot$ P(m_2<UCL$_A$)
   + [P(B) + P(B-)] $\cdot$ P(m_2<UCL$_B$)
   + [P(C) + P(C-)] $\cdot$ P(m_2<UCL$_C$)

where

$$P(m_2<UCL_A)=\int_{-\infty}^{Z_A}\emptyset(z)dz$$

$Z_A = (UCL_A - m_2)/ (s''/\sqrt{n_3})$
$UCL_A = m'' + 3 (s''/\sqrt{n_3})$
$m_2 = m'' + m (s''/\sqrt{n_1})$

$$P(m_2<UCL_B)=\int_{-\infty}^{Z_B}\emptyset(z)dz$$

$Z_B = (UCL_B - m_2)/ (s''/\sqrt{n_2})$
$UCL_B = m'' + 3 (s''/\sqrt{n_2})$
$m_2 = m'' + m (s''/\sqrt{n_1})$

$$P(m_2 < UCL_C) = \int_{-\infty}^{Z_C} \emptyset(z)dz$$

$$Z_C = (UCL_C - m_2)/ (\ s''/\sqrt{n_1}) = 3 - m$$
$$UCL_C = m'' + 3\ (\ s''/\sqrt{n_1})$$
$$m_2 = m'' + m\ (\ s''/\sqrt{n_1})$$

For example, for m=0, m=1 and m=3:

$$
\begin{aligned}
P_a(0) &= (.0428)(1) + (.2718)(1) + (.6826)(1) \\
&= .9986
\end{aligned}
$$

$$ARL(0) = 1/(1-P_a(0)) = 740$$

$$
\begin{aligned}
P_a(1) &= (.0428)(.90) + (.2718)(.94) + (.6826)(.98) \\
&= .96
\end{aligned}
$$

$$ARL(1) = 1/(1-P_a(1)) = 28.37$$

$$
\begin{aligned}
P_a(3) &= (.0428)(.01) + (.2718)(.11) + (.6826)(.50) \\
&= .37
\end{aligned}
$$

$$ARL(3) = 1/(1-P_a(3)) = 1.60$$

Economic   Analysis

As illustrated above there is a trade off between ARL reduction and ASN increase when the adaptation principle is applied to control charting. A model of this cost relationship is given below:

Assume the process is in control and capable, then the cost model generated by a shift in mean is given by:

Total Cost = Inspection Cost + Defectives Cost
Symbolically:

$$TC = C_i \cdot ARL \cdot n + C_f(m') \cdot ARL \cdot N \qquad (2)$$

where C_i is the inspection cost per unit and C_f is the failure cost per unit generated, if the process is producing units with quality characteristic at m'.

Since we assumed the process was capable, the Defective Cost before the shift would be zero. If we assume a quadratic cost model for the defectives generated by the process shift, we have:

$$C_f(x) = C(x-T)^2 \qquad (3)$$

where T is the target and C is a constant with dimensions \$ per specification dimension squared. Then,

$$C = C_f(UCL)/(UCL-T)^2 \qquad (4)$$

or by replacement

$$C_f(x) = (C_f(UCL) /(UCL-T)^2)\cdot(x-T)^2 \qquad (5)$$

or

$$C_f(m') = [((m'-T)/(UCL-T))^2]\cdot C_f(UCL). \qquad (6)$$

Hence, the mean shift cost relation becomes:

$$TC(m',n: N,T,s,C_i,C_f) = ARL\cdot(C_i\cdot n + C_f\cdot N) \qquad (7)$$

or

$$TC = ARL\cdot(C_i\cdot n + C_f(UCL)\cdot N\cdot((m'-T)/(UCL-T))^2) \qquad (8)$$

Figure 4 compares the mean shift total cost curves foraShewhart Mean control chart and an Adaptive Mean control chart.

Minimum Cost Plans

The mean shift total cost model is given by:

$$TC = ARL(C_i\cdot ASN + C_f(UCL)\cdot N\cdot((m'-T)/(UCL-T))^2) \qquad (9)$$

We know that ARL decreases as n increases and that ASN increases as n increases. The obvious question arises as to the

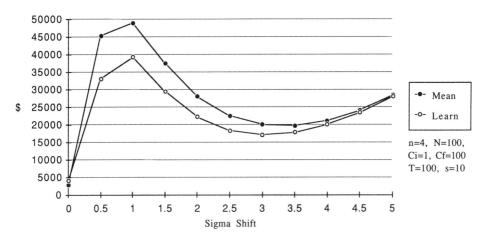

Figure 4. Total Cost of a Mean Shift

existence of minimum points of the total cost function. If we look at the TC function for different values of the sample size switch points Z_1 and Z_2, with fixed values of m, n, N, T, s, C_i, C_f, then the analysis indicates that a plan which is quick shifting (i.e., smaller Z_1) is less expensive for common ratios of n:N and $C_i:C_f$. Thus, the plan AM(1n,2n,3n), with switches at the standard values $Z_1=1$ and $Z_2=2$, generates more cost than does AM(1n,2n,3n) with switches at $Z_1=.5$ and $Z_2=2$.

However, if n is close to N and C_i is close to C_f, then the Total Cost function is not strictly decreasing over the interval $[0 \leq Z_1 \leq Z_2 \leq 3]$. For the examples reviewed, we still find that the absolute minimum occurs at small values of Z_1, but that the TC function then increases to a maximum at about $Z_1=1.5$ and $Z_2=2.5$ before it begins to decline.

Recommendations

For processes which often experience mean shifts in the one sigma range the use of an AM(1n,2n,3n) plan with switch points set at $Z_1=.5$ and $Z_2=2$ will provide significant savings over the same plan with switches set at $Z_1=1$ and $Z_2=2$.

It is possible to reduce TC even further by reducing Z_1 and Z_2, but the net cost improvement is less dramatic. Hence, we recommend using $Z_1 = .5$ and $Z_2 = 2$ for the switch points in $AM(n_1, n_2, n_3)$ plans. Still a simpler approach would be to use a two level plan, $AM(n_1, n_2)$ with single switch point set at $Z_1 = .5$. This plan has only slightly worse cost performance than a three level plan but is much simpler to apply.

Summary

The adaptation principle can be applied to any control charting scheme in which periodic sampling is being used.

Adaptive control charts can provide significant performance improvements over conventional control charts both in terms of average run length (ARL) and cost. They do require a computer for effective application, but the benefits should easily outweigh the cost.

Note: Adaptive control chart software is available from the author at Apple Computer, Inc.

8
MULTIVARIATE QUALITY CONTROL: 40 YEARS LATER

J. Edward Jackson
Consultant
Rochester, NY 14612

Introduction

Just forty years ago, a rather unique paper was published by Harold Hotelling entitled *Multivariate Quality Control* (Hotelling [1947]). This article was one of a number of papers combined into a volume summarizing some of the research done during World War II by the Statistical Research Group (Wallis [1980]). What was unique about this paper was that it was really the first one on the subject of multivariate quality control (MQC). Some of the concepts used went back to the turn of the century (Pearson, [1901]) and had been suggested by Shewhart [1931] but Hotelling's work (1) established a multivariate analog of the univariate control charts for means and dispersion, (2) included as much as was known at the time with regard to the relevant small sample distributions and (3) illustrated these techniques by means of an actual example involving the testing of bombsights. It is not the purpose here to discuss this paper in detail but rather to review what has happened in MQC in the forty years since and to present a summary of what is now available to the MQC practitioner.

Why MQC?

The basic Shewhart control chart for means (or individuals) is a hypothesis test that the current level of a process is equal to some standard value. Suppose the significance level for this test was a = .05. Then, if the usual normality assumptions held, one time in twenty this procedure would indicate a shift in the process when, in fact, no shift had occurred. Turned around, this says that 95% of the time when the process had not shifted, no action would be taken. If two such control charts, representing two different measurements on the same process, were being used simultaneously and the two variables were independent of each other, the probability that *neither* one would indicate an out-of-control situation when the process was on standard with regard to both variables would be (.95) (.95) = .9025 so the effective Type I error is about .10. If ten variables were employed, the effective Type I error would be

$$1 - (.95)^{10} = .40$$

so that ten separate control charts would be calling for action well more than half of the time when in fact none would be required.

One might be tempted to compensate for this by widening the limits to reduce the overall Type I error using techniques such as Bonferroni Bounds (Seber, [1984]) but this would have the undesirable effect of increasing the Type II error or the probability of failing to detect a shift in the process when one does occur. In addition, Bonferroni Bounds do not take into account any correlation which may exist between the variables.

If the variables are correlated, which is usually the case, one cannot make such simple statements about Type I and/or Type II errors. If the variables were perfectly correlated, one would only need to measure one of them and the Type I error would remain at .05. However, in reality, this will not be the case and with several related variables, all with different intercorrelations, determining the effective Type I error can be a major task. It is better to redesign the control chart methodology to take these correlated variables into account and, moreover, produce a single value with which to test the overall level of the process.   This is the T^2-test whose small-sample distribution Hotelling had derived in 1931 and which has the form:

$$T^2 = m[\bar{x} - std]'S^{-1}[\bar{x} - std] \tag{1}$$

where $\bar{x}$ is a vector of sample means based on a sample of size m, **std** is the corresponding vector of standards and S is the covariance matrix of these variables which in MQC is usually determined during a base period and is estimated with n degrees of freedom.   The distribution of T^2 is related to the F-distribution by the relation

$$T^2_{p,v} = \frac{p^{(v-1)}}{v-p} F_{p,v-p+1} \tag{2}$$

For details of this test along with many of the other techniques discussed in this paper, see Jackson [1985].   If m=1, this reduces to a control procedure for a vector of individual measurements rather than averages.   For the purposes of this paper, the situation for m=1 will be designated as T^2 and for m>1 as T_M^2.

Hotelling's Bombsight Paper

What Hotelling did in his 1947 paper was to introduce a set of *generalized* T^2-statistics.   The first of these T_0^2 represents the overall variation of a sample of observations from their standard.   If

$$T_i^2 = [x_i - std]'S^{-1} [x_i - std]$$

is computed for each observation in the sample, then

$$T_0^2 = \sum_{i-1}^{m} T_i^2 \tag{3}$$

This is a Lawley-Hotelling trace statistic for which tables may be found in Seber [1984] for large samples has an asymptotic chi-square distribution with p degrees of freedom. We have already shown (Equation 1) that we can compute another statistic, $T_M{}^2$, which represents the departure of the sample mean from its standard. Hotelling also developed a statistic, $T_D{}^2$, which represents the variability of the sample about its own mean. $T_D{}^2$ also has a Lawley-Hotelling distribution and can be calculated by the relationship:

$$T_D{}^2 = T_O{}^2 - T_M{}^2 \qquad (4)$$

In other words, the sum of the individual $T_i{}^{2}$'s, $T_O{}^2$ can be decomposed into two parts, $T_M{}^2$ representing the mean level and $T_D{}^2$ representing discussion. This is the multivariate analog of control charts for means and dispersion.

These techniques were illustrated by way of a two-variable example, the variables being the range and deflection deviations of the impact of a bomb from the target after allowance had been made for a number of explanatory variables such as direction of flight, bombsight model, etc. m would represent the number of bombs dropped at a single time. Hotelling also investigated the comparison of among and within flight variability.

After   1947

One should keep in mind that after 1947 statistical quality control was still considered a relatively new technique. Most people who used it did so in compliance with various aspects of government contract work during World War II. Once the war was over, some of these people could see applications of SQC to consumer products and took advantage of their wartime experiences. Others said, in essence, "Well, now that we aren't doing any more government contract work, we can get rid of all these control charts and sampling plans". Fortunately, there were enough of the former group to carry on a spirit of progress in SQC for another ten years or so, culminating in such innovations as CUSUM charts and geometric weighted moving average charts, as well as a consolidation of a myriad of inspection schemes into various MIL standards. Then, as we all know, SQC took a back seat to productivity for a couple of decades and is only now being "rediscovered".

But what of MQC? The world wasn't ready. There were not that many practicing SQC'ers in the first place and there were certainly fewer people versed in the basic philosophy, let alone the techniques of multivariate analysis. The intersection of these two groups of people was quite small. Many multivariate techniques had been developed but only in general terms. Much of the published material did not make use of matrix algebra, making it harder to read. Hotelling's work did stir up some interest however, particularly at the Eastman Kodak Company, where the generalized T^2-statistics were not only put into use on some photographic processes but in so doing, these techniques were often combined with another Hotelling technique, *principal components* (Jackson and Morris [1957]). With the exception of that and some work involving acceptance sampling, to be described in Section 9, things were fairly quiet for a number of years. But in time, MQC was also "rediscovered" and considerable effort was directed towards new and better techniques. The rest of this paper will be devoted to describing the various developments in MQC through the present time.

T^2-Control

The basic elements of T^2-control were already in place with the publication of Hotelling's 1947 paper. The distribution of T^2 had been derived by Hotelling in 1931 which, as we have noted, held both for testing of mean vectors and, assuming normality, vectors of individual observations. That is, this distribution is related to the F-distribution and hence it is in the hands of everyone. Furthermore, for power computations, the non-central F-distribution may be employed. The Lawley-Hotelling distribution, used for the distribution of T_O^2 and T_D^2, had been derived by Lawley [1938] and Hotelling [1941] and tabulated by Davis [1970a, 1970b, 1980] and then compiled in a single place by Seber [1984]. Some practitioners use, in place of T_D^2, the maximum likelihood ratio test for covariance matrices

(Anderson [1984]). Although this may be a more powerful test, MQC conditions for which it is more powerful have not been established. There is also the question of whether the increase in power is worth the extra labor involved in obtaining it. $T_D{}^2$ has the dual advantage of simplicity in both concept and computation (Equation 4).

Not much research has been done with cost functions in the multivariate case. However, Alt and Deutsch [1978] have developed a function related to the average run length. A more elaborate model has been developed by Montgomery and Klatt [1972a, 1972b] based on a number of different costs associated with monitoring the process as well as those related to producing defective product. See also Alt [1976].

MQC Using Principal Components

As mentioned in Section 4, the first use of principal component analysis (PCA) in quality control was due to Jackson and Morris (1957). Basically, PCA consists of transforming p correlated variables, x, into a new set of uncorrelated variables, y, called *principal components (pc's)* using the following relationship:

$$y = W'[x - \bar{x}] \tag{5}$$

where W' is the transformation matrix made up of *characteristic vectors*. Examination of the coefficients of these vectors may require some interpretation as to the meaning of the pc's.

The method of principal components is due primarily to Hotelling [1933] although the original concept goes back to Karl Pearson [1901]. In many industrial applications, the pc's do have physical interpretation and can be used as control variables in their own right. The same generalized T^2-statistics may still be employed and in the case of an indication of an out-of-control situation, the diagnosis of this condition may be enhanced by virtue of the fact that the pc's are uncorrelated. For a description of the state-of-the-art of PCA, see Jackson [1980, 1981a, 1981b].

One of the features of PCA is that a reduced number of pc's, $k<p$ may adequately describe the variability of the original x and if that is the case, the work of the control personnel may be lessened considerably. However, in that case, an additional control should be maintained to the extent which the retained pc's do not predict the original variables. Two techniques have been proposed. The first, due to Jackson and Morris, involves the sum of squares of deviations between the original data and the values of these variables as predicted by the k pc's. The distribution of this statistic was derived by Jackson and Mudholkar [1979], who also produced some generalizations similar to the generalized T^2-statistics. The second technique involves the use of the $p-k$ unretained pc's (Rao [1964]; Hawkins [1974], [1980]) or the use of the largest unretained pc (Hawkins [1974], Fellegi [1975]).

One of the larger applications of these PCA techniques dealt with the monitoring of audiometric examinations of a large number of employees. While not related to a product or process, it involved a comparison of data with a set of reference values and hence made use of all these control procedures in addition to some new ones, required to deal with allowance for age and to control absolute values of pc's (Jackson and Hearne [1978], [1979]).

Multivariate CUSUM

Presently most activity in MQC appears to be related to CUSUM charts. Much of this work is unpublished. CUSUM charts have always been popular because of their relatively shorter average run lengths for a large class of situations. With the resurgence of statistical quality control, there have been accompanying new developments in CUSUM charts, some of it spilling over into MQC. The first of these (Woodall and Ncube [1984]) consisted of maintaining a CUSUM chart for each variable and concluding the process to be out-of-control as any one of the CUSUMs exceeded its limit. This technique did not take into account the correlations among the variables. However, Woodall and Ncube also proposed the same scheme with principal components, which resulted in considerably shorter average run lengths, particularly when the intercorrelations among the original variables were large.

Two more proposals are the result of work by Pignatiello and Kasunic [1975], Healy [1987], Crosier [1988] and some unpublished work of Pignatiello. The first of these consists of computing T^2 for each observation and using these to produce a CUSUM of T^2. This does reduce the problem to a single statistic but does not take the *direction* of change into account. The second scheme essentially obtains a CUSM for each variable and obtains a T^2-statistic for these. This second technique appears to be the most promising of the ones proposed so far but it is still too early to tell. More experience is needed. Healy also gave a multivariate scheme for detecting an increase in variability.

A procedure similar to CUSUM is the *geometric weighted moving average*, sometimes known as *exponential smoothing*. Some multivariate extensions of this are currently being investigated by S.E. Rigdon and C.W. Champ.

Andrews Plots

A popular technique for multivariate data analysis is the Andrews plot (Andrews [1972]). This consists of a linear transformation of the original p variables into a single new variable of either the form:

$$f_1(\theta) = x_1/\sqrt{2} = x_2\sin(\theta) + x_3\cos(\theta) + x_4\sin(2\theta) + x_5\cos(2\theta) + \ldots \quad (6)$$

or

$$f_2(\theta) = x_1\sin(\theta) + x_2\cos(\theta) + x_3\sin(2\theta) + \ldots \quad (7)$$

depending on where p is odd or even. θ is defined over the range:

$$-P \leq q \leq P$$

so that this new transformed variable is actually a continuous function over that range. These plots have been used primarily as a graphical clustering technique in that data which have similar characteristics will having similar curves. In theory, at least, the shapes of these curves lend themselves to interpretation; a simple example may be found in Jackson [1985].

The use of Andrews plots as a control tool has been proposed by Kulkarni and Paranjape [1984, 1986] who have constructed control limits for these plots based on the covariance matrix of the original variables. As in the case of the multivariate CUSUM, this technique is also very new and it will be some time before its strengths and weaknesses will have been ascertained. What is encouraging is the sudden burst of new ideas in the last three or four years.

Multivariate Acceptance Sampling

In contrast to control charts, the technique of *acceptance sampling* involves sampling a number of items from a lot of finished product to determine whether or not the quality of this lot is such that it should be shipped. ("Finished" in this case refers to any operation at the conclusion of which the product is shipped somewhere else, be it to the ultimate consumer or merely another department in the same organization.)

In the case of sampling for variables, specifications are made on a characteristic of the product and a sampling scheme is created by establishing a Type I error associated with the standard for the product and a Type II error associated with the specification limits. This will determine the sample size required and the limits used to make the required decision about the lot. This differs from control charts in that the Type II error is rarely set up for specification limits in advance with the exception of acceptance control charts which are discussed in the next section. As was the case for control charts, great strides were made in acceptance sampling during World War II, with much of the development work being sponsored by the government, from which came the beginning of the various MIL STD documents still in use today.

While there has not been a lot of activity related to multivariate control charts, even less is available for multivariate acceptance sampling. The reason for this is that the specifications for the product will generally consist of a collection of individual specifications for each variable. The intersection of these specifications would form a rectangular solid while the variability would more than likely take the form of an ellipsoid. A number of possibilities may be suggested:

1. Bonferroni Bounds These would have the same properties as in their use for control charts.   They would have relatively poor power and would not take into account the relationships among the variables. They would, however, be easy to understand and administrate. Chapman et al. [1978] have proposed something similar to this in which a sample from a lot is inspected one variable at a time and, essentially, rejection on the basis of a single variable rejects the lot.   Their model takes into account costs of inspection, screening, repair and the Type I and Type II errors.   The model also takes into consideration whether: (1) the inspection process is totally destructive on the basis of a single variable, (2) the inspected item may be subjected to tests on all variable but is then no longer fit for sale or (3) the procedure is nondestructive and the item may be sold.

2. Use the philosophy of acceptance control charts to establish a "compromise" ellipse associated with the specifications and base a multivariate test on this.   This may be somewhat better than the Bonferroni limits but would be ad hoc and would not take into account the correlations among the variables.

3. Base the sampling plan on some relationship of the specifications which take the form of a rectangular solid and the inherent variability which would take into account the correlations among the variables. Jackson and Bradley [1961] considered this problem for some sequential sampling procedures.   Their example consisted of a three-variable problem, the variables being related to various characteristics of the performance of a ballistic missile.   They tried both inscribing within and circumscribing about this rectangular solid ellipsoids proportional to the inherent variability.   The former procedure would be too stringent; the latter, too lenient. Presumably, some optimum might be obtained, again possibly in the spirit of acceptance control charts but no work has been done along this line to date.   Baillie [1986] has proposed a scheme based on the proportion of the lot that would conform to the specifications.

4. Ellipsoidal specifications.   Shakun [1965] considered
specification which was also in the form of ellipsoids
which had the effect of reducing both the specifications
and the test statistic to quadratic forms upon which the
test is based.

5. Principal components.   Jackson and Bradley [1966] later
suggested using principal components in which case the
specifications could be made on the principal components
themselves.   This would then allow both the specifications
and the inherent variability to be expressed as quadratic
forms and might be easier to perform, particularly for the
specifications, providing the principal components were
easily interpretable and readily understood by the people
responsible for setting the specifications.   For example,
this technique might be useful if PCA had already been
used for controlling the processes which produced the
items being inspected.   However, the use of acceptance
control charts might allow the user to perform both
functions.

Note that not much new work has been done in this field
in the past twenty years and that this field should present a
great opportunity for research and development of better
procedures.

Acceptance   Control   Charts

Acceptance   Control   Charts (Freund [1960];   Schilling
[1982]) use essentially the same philosophy in setting up
control charts that one does in setting up acceptance sampling
plans.   The main difference between these charts and Shewhart
control charts is that the Shewhart chart is testing for
departures from standard while acceptance control charts are
concerned with the proportion of material failing to conform to
specifications and are employed when the specifications are
wider than the limits based on the variability of the process.   To
extend acceptance control charts to the multivariate case would
involve all of the same problems outlined in the previous
section.   In addition, there is now some controversy over this
technique since it may allow for some departure of a process

level from its standard and those who belong to the "zero
defect" school would argue against its use. However, proponents
of acceptance control charts would be quick to point out that
what makes a product competitive in the long run is cost which
could be reduced by using these charts, yet keeping the
outgoing product within specifications subject to the specified
Type II error.

What the Future Holds

In view of the interest in statistical quality control and
the increased facilities available to incorporate multivariate
procedures into it, the future should provide an exciting
atmosphere for research and development in these activities.

As discussed in Section 9, the area of multivariate
acceptance sampling is virtually undeveloped. The main effort
here should be directed towards the construction of
multivariate specifications which can be more readily translated
into sampling plans; once this is done, the actual multivariate
tests required will probably require the use of statistics which
have already been developed.

In the area of process control, the field is just as
unexplored. There has been no work done on multivariate
acceptance control charts because it involves some of the same
problems as acceptance sampling itself. Many of the techniques
presently in use involve large-sample distributions. While
these may be adequate for most QC applications, it would be
helpful to have the exact distributions for the others. The
problem of cost functions is another fertile field. These are just
a few of the many problems still to be researched.

Most important, for the moment, however, is application
of the methods already available to real life situations. This
should produce more quality control engineers who can deal
with the concepts of multivariate analysis and at the same time
produce more multivariate-minded statisticians who are
knowledgeable about the fundamentals of quality control.
When this happens, R&D teams consisting of both groups of
people will be formed to further the exploration of this
important topic.

References

Alt, F.B. (1976) Corrections to papers by Montgomery and Klatt, Management Sci.,22,, 1167-1168.

Alt, F.B. and Deutsch, S.J. (1978) "Multivariate economic control charts for the mean", pro. Seventh Ann. Meeting NE Reg. Conf. Amer. Inst. for Dec. Sci. June 1978, 109-112.

Anderson, T.W. (1984) An Introduction to Multivariate Analysis, 2nd Ed., John Wiley and Sons, Inc., New York.

Andrews, D.F. (1972) Plots of high-dimensional data, Biometrics, 28, 125-136.

Baillie, B.H. (1986) "Multivariate acceptance sampling", Third International Workshop on Statistical Quality Control, Copenhagen, Denmark.

Chapman, S.C., Schmidt, J.W. and Bennett, G.K. (1978) "The optimum design of multivariate acceptance sampling plans", Naval Res. Logist. Q., 25, 633-651.

Crosier, R.B. (1988) Multivariate generalizations of cumulative sum quality-control schemes, Technometrics, 30, 291-303.

Davis, A.W. (1970a) "Exact distribution of Hotelling's generalized T_0^2", Biometrika, 57, 187-191.Correction: 59 [1972], 498.

Davis, A.W. (1970b) "Further applications of a differential equation for Hotelling's generalized T_0^2", Ann. Inst. Stat. Math., 22, 77-87.

Davis, A.W. (1980) "Further tabulation of Hotelling's generalized T_0^2", Comm. Stat. Simul. Comput. B, 9, 321-336.

Fellegi, I.P. (1975) "Automatic editing and imputing of quantitative data", Bull. Int. Stat. Inst., 46, 249-253.

Freund, R.A. (1960) "A Reconsideration of the variables control chart with special reference to the chemical industries", Ind. Qual. Cont. 16, No. 11, May 1960, 35-41.

Hawkins, D.M. (1974) "The detection of errors in multivariate data using principal components", J. Amer. Stat. Assoc.,69, 340-344.

Hawkins, D.M. (1980) Identification of Outliers, Chapman and Hall, London.

Healy, J.D. (1987) "A note on multivariate CUSUM procedures", Technometrics, 29, 409-412.

Hotelling, H. (1931) "The generalization of student's ratio", Ann. Math. Stat., 2, 360-378.

Hotelling, H. (1933) "Analysis of a complex of statistical variables into principal components", J. Educ. Psych., 24, 417-441, 489-520.

Hotelling, H. (1947) Multivariate quality control, Techniques of Statistical Analysis, Eisenhart, Hastay and Wallis, Editors, McGraw-Hill, 111-184.

Hotelling, H. (1951) "A generalized T test and measure of multivariate dispersion", Proc. 2nd Berkeley Symp.on Math. Stat. and Prob., Univ. of California Press, Berkeley, Cal., 23-41.

Jackson, J.E. (1980) "Principal components and factor analysis: Part I- Principal components", J. Qual. Tech., 12, 201-213.

Jackson, J.E. (1981a) "Principal components and factor analysis: Part II- Additional topics related to principal components", J. Qual. Tech., 13, 46-58.

Jackson, J.E. (1981b) "Principal components and factor analysis: Part III- What is factor analysis?" J. Qual. Tech., 13, 125-130.

Jackson, J.E. (1985) "Multivariate quality control", comm.,stat. Theor. Meth.,14, 2657-2688.

Jackson, J.E. and Bradley, R.A. (1961) "Sequential x^2- and T^2- Tests and their application to an acceptance sampling problem", Technometrics 3, 519-534.

Jackson, J.E. and Bradley, R.A. (1966) "Sequential multivariate procedures with quality control applications", Multivariate Analysis, I, P.R. , Ed., Academic Press, 507-519.

Jackson, J.E. and Hearne, F.T. (1978) "Allowance of age in the multivariate analysis of hearing loss", Biometrie-Praximetrie, 18, 83-104.

Jackson, J.E. and Hearne, F.T. (1979) "Hotelling's T_M^2 for principal components - what about absolute values?" Technometrics, 21, 253-255.

Jackson, J.E. and Morris, R.H. (1957) "An application of multivariate quality control to photographic processing", J. Amer. Stat. Assoc., 52, 186-199.

Jackson, J.E. and Mudholkar, G.G. (1979) "Control procedures for residuals associated with principal components", Technometrics, 21, 341-349.

Kulkarni, S.R. and Paranjape, S.R. (1984) "Use of Andrew's function plot technique to construct control curves for multivariate process". Comm. Stat. - Theory and Methods, 13, 2511-2533.

Kulkarni, S.R. and Paranjape, S.R. (1986) "An improved graphical procedure for multivariate quality control". Comm. Stat. - Sim. and Comp., 15, 135-146.

Lawley, D.N. (1938) "A generalization of Fisher's z-test", Biometrika, 30, 180-187.

Montgomery, D.C. and Klatt, P.J. (1972a) "Economic design of T_2 control charts to maintain current control of a process", Man. Sci., 19, 76-89.

Montgomery, D.C. and Klatt, P.J. (1972b) "Minimum cost multivariate quality control tests", AIIE Transactions, 4, 103-110.

Pearson, K. (1901) "On lines and planes of closest fit to systems of points in space", Philosophical Magazine, Series 6, 2, 559-572.

Pignatiello, J.J. Jr. and Kasunic, M.D. (1985) "Development of a multivariate CUSUM chart", Proc. 1985 ASME Int. Computuers in Engineering Conf. and Exh., Boston, Mass.

Rao, C.R. (1964) "The use and interpretation of principal component analysis in applied Research", Sankha, Ser.A., 26 329-358.

Schilling, E.G. (1982) Acceptance Sampling in Quality Control , Marcel Dekker, 169-173.

Seber, G.A.F. (1984) Multivariate Observations, John Wiley and Sons, Inc., New York.

Shakun, M.F. (1965) "Multivariate acceptance sampling procedures for general specification ellipsoids". J. Amer. Stat. Assn., 60, 905-913.

Shewhart, W.A. (1931) Economic Control of Quality of Manufactured Product, D. Van Nostrand Co., Inc. New York.

Wallis, W.A. (1980) "The Statistical Research Group 1942-1945 (with discussion)", J. Amer. Stat. Assoc., 75, 320-335.

Woodall, W.H. and Ncube, M.M. (1985) "Multivariate CUSUM quality control procedures". Technometrics, 27, 285-292.

SECTION II

EXPERIMENTAL DESIGN FOR PRODUCT AND PROCESS DEVELOPMENT

9
DESIGN GOALS: MATHEMATICAL AND STATISTICAL IMPLICATIONS

J. Bert Keats

Director, Statistical and Engineering Appplications
for Quality Laboratory
Computer Integrated Manufacturing Research Center
College of Engineering and Applied Sciences
Arizona State University
Tempe, AZ 85287-5106

Introduction

With the recent, and long-anticipated, emphasis on achieving quality and reliability principally through design activities comes numerical targets and goals. Many companies now have the unfortunate experience of having a goal without a means of identifying whether or not the goal has been reached. Additionally, companies are discovering that achieving target yields from manufacturing processes may not be cost-effective if the means of achieving them is through extensiveinspection or testing. Another problem which existed before the prevailing design emphases and which is now greatly magnified is the notion of setting a goal as the mean of a random variable without taking distributional properties under consideration.

This chapter addresses each of these points and provides suggestions for improvement.

DPM/PPM Goals

The Time Between Events CUSUM. Companies setting fraction defective goals in terms of defects per million (DPM) or equivalently parts (defective) per million (PPM) are usually striving for continuous improvements. As technology and quality practices improve, target values are becoming smaller and smaller. 100 parts per million defective is a common goal in many industries, whereas PPM levels below 50 exist in a few others. It makes no sense to set such lofty goals without sound procedures for identifying whether or not such goals are being achieved. Furthermore, one should be able to identify shifts or changes (either up or down) in the small fraction defective rate.

It is obvious that when defect levels are of the order of magnitude mentioned above, sampling is of no use as sample sizes large enough to allow the possibility of a single defective would approximate 100 percent inspection. Lucas [1985] has provided the methodology for dealing with small fraction defectives. Suppose that it is possible to obtain a rough count of the number of non-defective items produced between defectives. These counts may be obtained by combining information about defectives found during manufacturing with those items found to be defective in the field. It is obvious that record-keeping must be such that time order of production is roughly preserved so that counts of good items occurring between defectives are reasonably accurate. The Time Between Events (TBE) CUSUM (increasing rate case) is given by

$$S_i = MAX (0, k-y_i + S_{i-1}) \qquad (1)$$

where S_i is the CUSUM value after the ith defective item is found, k is a constant based on the design of the plan, y_i is the number of good units between the i-1st and ith defective and S_{i-1} is the previous CUSUM value (value after the i-1st defective was found). To find k, one must first specify acceptable and detectable defect rates, u_a and u_d, respectively. Suppose, for example, that the target value was 50 PPM and 100 PPM is deemed unacceptable. Then $u_a = 50$ and $u_d = 100$. We next compute

$$k_t = (\ln u_d - \ln u_d)/ (u_d - u_a) * u_a \qquad (2)$$

In our example, $k_t = \ln 2/(2u_a - u_a) * u_a = \ln 2 = 0.693$. In fact, any ratio $u_d/u_a = 2$ will produce a k_t ratio of 0.693. Next, we use the Average Run Length Table provided by Lucas, a portion of which is shown in Table 1. The purpose of the table is to select h_t, which will determine the control limit for the plan.

The basis for selection is the Average Run Length (ARL) which is defined as the average number of observations which have occurred after the shift, but before an out of control signal is given. Two columns of the ARL table are used; the column which indicates no shift (count rate x $u_a = 1$) and the column representing the shift to be detected (count rate x $u_a = 2$, in this case). Clearly, we desire a large ARL when no change has taken place and a small ARL when the detectable change occurs. The choice of h_t is thus very subjective and is strictly a function of the decision-maker's perceptions of "large" and "small". Obviously, these perceptions are related to the cost of searching for a false alarm and the cost of allowing defects to escape. For our purposes, suppose that $h_t = 2.8$ is selected. To complete the plan, we now consider the magnitude of u_a. $k = k_t/u_a = 0.7/50/10^6 = 14,000$ and h, the upper control limit $= h_t/u_a =$

Table 1. Average Run Lengths : Time Between Events CUSUM, increasing rate case. No FIR Feature

h_t	k_t	h_t/k_t	Count Rate as a Multiple of Acceptable Rate, m_t.	
			1.0	2.0
2.1	.7	3	49.9	9.56
2.8	.7	4	110.	12.9
3.5	.7	5	230.	16.4
4.2	.7	6	468.	19.9
4.9	.7	7	948.	23.4
5.6	.7	8	1870.	26.9

From Lucas, J.M. (1985), "Counted Data CUSUM's", Technometrics, Vol. 27, No. 2, Table 7, p. 136.

$2.8/50/10^6 = 56,000$. Note that k provides a "buffer" around the observed good items between defectives. When the process is meeting the target, the expected value of y_i is 20,000, the reciprocal of u_a. The CUSUM will not increase until the observed value of y_i exceeds 14,000. We will conclude that the fraction defective rate has shifted to 100 PPM if and when the CUSUM exceeds 56,000. The k_t value is appropriate for <u>any</u> detectable to acceptable mean defective ratio of 2. Likewise, for h_t. One has only to change the k and h values. For example, if u_a = 2 PPM and u_d = 4 PPM, then $k = 0.7/2/10^6 = 350,000$ and $h = 2.8/2/10^6 = 1,400,000$.

When the u_d/u_a ratio is 1.5, k_t is approximately 0.8. The increasing rate case has been illustrated here. For the decreasing rate case, the appropriate CUSUM is

$$S_i = \text{MAX} (0, y_i\text{-}k + S_{i\text{-}1}) \qquad (3)$$

All determinations and calculations are analogous to the increasing rate case. That is, the detectable rate is smaller than the acceptable rate and S_i only increases for "larger" times between defects, which is indicative of a decreasing failure rate. Operating both CUSUM's simultaneously constitutes a two- sided test for shifts in defect levels.

<u>Research with the Time Between Events CUSUM</u>. Using computer simulation of a Poisson process, we have investigated the behavior of the TBE CUSUM under a variety of conditions. Figure 1 below illustrates three types of shift conditions used in the simulations (continuous, sudden and step shift). Note that the defect rate was altered as a function of cumulative defects, not trials (numbers of good and bad items manufactured). For purposes of identifying false alarms (CUSUM test indicated a change, yet no change existed) 1.25 times the acceptable defect rate was chosen as the point below which out-of-control signals were counted as false alarms. A simulation run terminated either when the test indicated out-of-control (CUSUM value above h) or when 80, 60 or 70 (counts for the continuous, sudden and step shifts, respectively) defects had been produced without an out- of-control signal (a miss or Type II error).

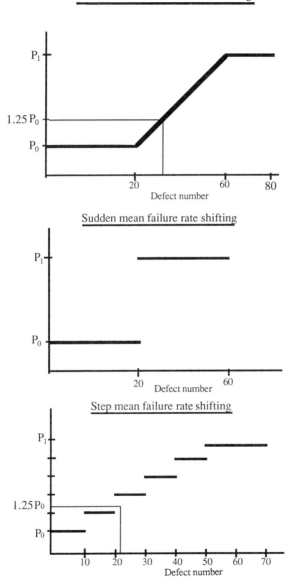

Figure 1. Shift Conditions used in the Simulations

Table 2 presents the simulation results for a variety of h values. It may be observed that the test did very well, averaging 91 percent correct decisions. Selection of an h_t and hence an h value tempers the false alarms and misses. That is, for a fixed value of k_t (determined by u_a and u_d), smaller h_t values tend to give more false alarms. This implies that larger h_t values yield more misses and a longer wait for detection. We have studied average and earliest detection points for a variety of h values across the three shift conditions for each of three hypothetical manufacturing problems. The latest detection point is always a miss. For the continuous shift the average detection point was at the 46th defect. For the sudden shift it was at the 39th defect. For the step shift, the average detection point was at the 40th defect produced. The earliest detection point was about 9 for all conditions. This of course, is a false alarm. Lucas suggests a Fast Initial Response (FIR) CUSUM for those cases in which it is possible that the system may start at an out-of-control condition or whenever a shift to the detectable state is possible in early trials. He gives ARL tables for the FIR case so that h_t may be selected. For the FIR, the CUSUM, rather than starting at zero is given a "head-start". The value used for the head start (S_0) is half the value of h, the CUSUM control limit.

Table 2.   TBE CUSUM Results (4500 Simulation Runs for each Type of Shift)

μa = 50 ppm, μ_d = 100 ppm;   variety of h values; ( ) percent

SHIFT	CD	FA	M
Continuous	4025 (89)	319 (7)	156 (4)
Sudden	4181 (93)	144 (3)	175 (4)
Step	4044 (90)	206 (5)	250 (5)
Total	12550 (91)	669 (5)	581 (4)

CD = Correct Decisions;   FA = False Alarms;   M = Misses

The question of robustness arises in situations such as that described above in which simulated data generates variates satisfying the exact requirements of the test. The test is based on a Poisson process (exponential time between defects) and it was with exponential data that the TBE CUSUM was exercised. Figure 2 depicts how means for the three shift conditions were chosen. The presence of a defect was determined by selection of a random number. For example, if the mean defect rate was to be 5,000 ppm, an item was judged to be defective if the associated random number was between 0 and 0.005. Otherwise, it was considered a good item.

Table 3 presents a comparison of the CUSUM's efficiency for the non-Poisson cases described above as well as for the Poisson case. For the Poisson, the same mean defect rate was used as for the Poisson, and the same random number stream was used to generate defectives. Although 500 runs is not a large number, there is strong evidence to conclude that the TBE CUSUM is extremely robust. Two other experiments added more validity to the robustness claim. In the first experiment, a quality engineer with 4 years of experience in trouble-shooting problems in the manufacture of printed circuit boards constructed 32 scenarios representing actual problems (in the form of defect sequences) observed on the manufacturing floor. The TBE CUSUM was successful in identifying the problem in 31 of the cases. In the second experiment, the ability of the

Table 3. TBE CUSUM Result (Poisson/Non-Poisson Data)

500 Simulation Runs for Each Type of Shift
μ_a = 50 ppm, μ_d = 100 ppm

SHIFT	CD	FA	M
Continuous	416/412	84/87	0/1
Sudden	466/486	18/9	16/5
Step	468/472	15/12	17/16
Total	90/91	8/7	2/2

CD = Correct Decisions; FA = False Alarms; M = Misses

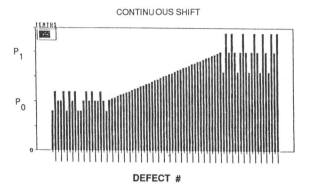

CONTINUOUS SHIFT

DEFECT #

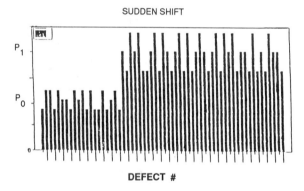

SUDDEN SHIFT

DEFECT #

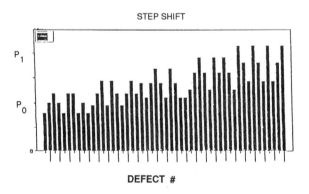

STEP SHIFT

DEFECT #

Figure 2. Shifts using Non-Poisson Data

CUSUM to function when defects within a lot are scrambled (out-of-sequence) as they might be in a production situation was tested for lot sizes of 15 and 50.   Whenever a defect was found within a lot, it was randomly assigned to one of the 15 or 50 positions in the lot rather than preserving its order.   Results indicated   no decrement in the CUSUM test's ability to make correct decisions in the presence of rearranged defects within the lots.

Use of an Expert System.   Much has been written about the use of an expert system in diagnosing a fault and recommending corrective action.   Often such suggestions for the use of this type of artificial intelligence are accompanied with reports of successful implementations.   However, as of this writing, there have been virtually no citations in either the expert systems or statistical literature which focus on the application of an expert system to a statistical test or on the data being used for such a test.   It is clear from conversations with several statisticians and quality engineers that research and implementations in this area have been done; the lack of published accounts is probably due to the advantage to be gained by keeping such developments   proprietary.

    In the majority of instances, the operator of a process station lacks both the time and expertise to define and reason with the procedures, the data and the statistics calculated from the data.   This highlights the potential need for a knowledge-based system to be integrated with the procedures to provide reason capabilities in a timely manner.   The simulation research described in this section was part of a project with Digital Equipment Corporation.   The following suggestions concerning the use of expert systems were made by Keats, Date and Kim (1988) in their final report: (1) the expert system is used to select parameters for the statistical tests.   It combines history, analysis and intuition and uses trade-offs (e.g., false alarms vs. misses) to select parameters such as the magnitude of the shift to be detected, the head start or FIR and the margin of error, (2) the expert system applies deep and hueristic knowledge for making decisions concerning : (a) the extent to which past data should be used in current situations, (b) the "weights" or values to be used when past and present data are combined for decision-making purposes and (c) the form (detail or summary)

in which data is to be stored for future use, (3) the expert
system may dictate certain "warm-up" periods for tests known
to take time before any decision is made (such as Wald's
Sequential Probability Ratio Test), (4) the expert system may be
used to compare current process data with past data to
determine if target values should be changed to reflect process
improvements, and (5) when two or more statistical tests are
applied to the same set of data, the expert system is used to
decide whether or not an undesirable shift has occurred--not by
applying voting rules, but rather by allowing the tests to work
in "concert" with each other. For example, suppose that one test
has generated an "out-of-control" signal. The expert system
would use information about how "far along" the other tests are
in yielding the same result and base its conclusions accordingly.

Conclusions. The TBE CUSUM is at least 90 percent accurate and
extremely robost. It represents a new tool for monitoring
extremely small defect rates whenever there is a means to
approximate the number of "good" items produced between
defects. The accuracy of this test could be improved by
reducing either misses or false alarms or both. Since selection
of the h_t parameter can influence false alarms and misses, one
could set h_t lower if there was a way of dealing with false
alarms. Misses would no longer be a problem. One way of
dealing with false alarms is to use a weighted CUSUM wherein
defects found during early trials do not receive as high a weight
as those occurring later. This assumes that the process starts
"in-control". Another suggestion is to use an expert system
either alone or in tandem with a weighted CUSUM to decide the
extent to which early defects should contribute to the CUSUM
value.

First Time and Other Yield Goals

System designers in companies practicing Design for
Manufacturability (DFM) are taught to exercise judgement in
specifying test and inspection stations through participation in
scenarios representing material flow through such stations (see
Figure 3). The designers are provided with knowledge that
First Time Yield (FTY), the percentage of product which is
assembled and shipped without rework or scrap (Ishikawa

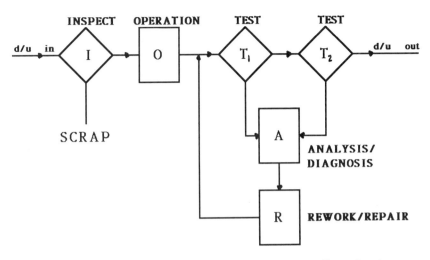

Figure 3. Material Flow through Inspection and Test Stations

(1985) calls this the "go-straight percentage") is the product of the Poisson probabilities of no defects (zero defects per unit or zero d/u) in each of the manufacturing operations which present opportunities for defects. Hence, FTY = exp $(-d_i/u_i)$. Yield at any operation is exp (-d/u). Other relevant calculations include Manufacturing Starts at any operation = 1/Yield, Total Test Time Per Unit = Starts x test time per unit for non-repairable units, Total Test Time Per Unit = (1 + d/u) x test time per unit for repairable units, Capacity = 1/(Total Test Time Per Unit) and Total Inspection Time Per Unit = 1 + (1 - Yield) = 2 - exp (-d/u). Exercises with scenarios such as that depicted in Figure 3 involve specifying an input d/u and a test effectiveness (probability of detecting a defect) as well as providing estimates of the cost of inspection, cost of test, cost of analysis, cost of rework, etc. . In Figure 3, "I" is and inspection station (perfect inspection assumed), "O" is a manufacturing operation, "T_1" and "T_2" are test stations with less than 100 percent effectiveness, "A" is an analysis or diagnosis station and "R" represents rework or repair. The system designer then performs calculations for scenarios with various configurations of test and inspection equipment to ascertain the effect on yield, cycle time, capacity and cost. It soon becomes apparent that with more inspection and test, the d/u at output decreases (yield increases), but cycle time and cost are driven upward and capacity decreases. The lesson to be learned is that achieving better yields by extensive use of test and inspection equipment is both time consuming and expensive. The obvious and optimal alternative is to design and manufacture better parts so that the need for test and inspection is greatly reduced.

Reporting and Comparing Small Fraction Defective Levels

Care must be exercised in comparing yields and other statistics based on PPM levels. Consider three different types of printed circuit boards, each having the same fraction defective levels, expressed in PPM. Suppose further that defects due to problems with vendors, automatic insertion of parts and the wave solder process have been identified. These rates are shown in Table 4. A yield (exp (-d/u)) calculation is performed for each type of board as seen in Table 5. Note that the only differences among the three types of boards is in the number of components on each board. Yet, the yields are vastly different. When comparisons like this are made, a penalty has been imposed on the boards with more components. To avoid this situation, yield should not be computed for comparison purposes. When comparing two or more units, calculate PPM/part which is given by PPM/part = defects/board x 10^6 parts/unit. For each of the three boards, this figure would be 3500 PPM/part and each type of board would be judged to

Table 4. Defect Rates (Printed Circuit Board)

	Vendor Problem (V)	Machine Ins. (M)	Wave Solder (W)
PPM	200	900	800
d/u	.0002	.0009	.0008

Each board averages 3 solder joints/part.

Table 5. Field Comparisons : Three Boards with Varying
Parts/Board

Board	Nbr. Parts	Defects per Unit				
		V	M	W	Total	Yield (%)
1	100	.02	.09	.24	.35	70.5
2	500	.10	.45	1.2	1.75	17.4
3	1000	.20	.90	2.4	3.5	3.0

have the same level of defect problem, as they should be. Some companies use "defects per million opportunities (DPMO)" or parts (defective) per million opportunities (PPMO) as the metric for comparisons. An opportunity is an occasion for a defect to occur. For example, on a printed circuit board with 200 components there may be 1700 opportunities for a defect to occur--i.e., each solder joint is an opportunity, each insertion of a component is an opportunity, etc.

Reliability and Maintainability Design Goals

Many of the activities associated with efforts to achieve quality during the early stages of product and process development now involve reliability and maintainability as well. It is most unfortunate that when reliability and maintainability goals are set and when metrics are defined for both the product and the equipment used in the processes associated with product manufacture, such goals and associated metrics are often expressed in terms of means--e.g., mean time to failure must exceed 1000 hours or mean time to repair must not exceed 30 minutes. Inputs to shop floor scheduling algorithms and software which predicts yields and cycle times likewise are often based on reliability and maintainability means. What is ignored by those who focus on the mean in these activities is the intelligent use of the properties of the underlying distribution. Consider the use of the exponential distribution for time between failures (Figure 4). It is an easy

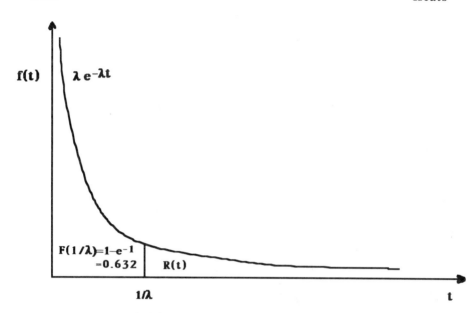

Figure 4. Reliability : Exponential Case

task to show that 63.2 percent of the failures will occur before the mean.   Surely we don't want to design a device or a system or specify cycle times based on an event (failure at the mean) for which most of the activities will have already occurred. What is more meaningful is to specify a time (or number of cycles) such that a specified large percentage of devices will still be operating at that time--e.g., design for an exponential failure distribution so that 95% reliability is achieved at 1000 hours. In this case, the mean can also be specified. Since R(t)= exp(-λt), λ= -ln R(t)/t. For 95 % reliability at 1000 hours, λ=-ln(.95)/1000 = .0000513 failures/hour, and the MTBF = 1/λ = 19,296 hours.

For the Weibull distribution, which also enjoys much use in reliability studies, goals and metrics set for the mean also have little validity.   With a two parameter Weibull distribution, and a shape parameter greater than or equal to one (which is almost always true), the mean will always be less than or equal to the 63.2   percentage point.   Table 6 presents the Weibull density, cumulative density and reliability function followed by log transformations which make the characteristic value, theta, a linear function of the inverse of the shape parameter, beta.

Table 6.   Reliability : Weibull Case

$$f(t) = (bt^{b-1}/q^b)\exp[-(t/q)^b]$$

$$F(t) = 1-\exp[-(t/q)^b]$$

$$R(t) = \exp[-(t/q)^b]$$

$$-\ln R(t) = (t/q)^b$$

$$-\ln(\ln R(t))=b\ln t-b\ln q$$

$$\ln q=\ln t-\ln(-\ln R(t))(1/b)$$

These two parameters completely define the distribution and they can be estimated by specifying a pair of reliability goals.

Suppose that design specifications are stated such that 90% survivability is desired for t_1 hours and 10% survivability is desired for t_2 hours, where t_2 is much larger than t_1 (see Figure 5).   Either graphically or analytically, the intersection

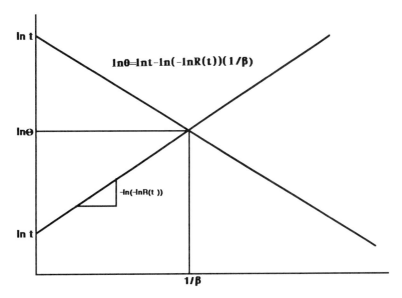

Figure 5. Suggested Reliability Goals : Weibull Case

point of the two lines with slopes -0.834 and +2.250 can be determined. The result is a theta and a beta value which satisfies the reliability goal and which can be estimated by the designer of the device or equipment.

For the maintainability case, the most popular distribution to describe both repair time and down time is the lognormal. Figure 6 presents a typical lognormal distribution. The median is always greater than the mode and the mean is always greater than the median.

Information associated with cumulative areas of the lognormal is easily obtained using the normal distribution, since the lognormal variate is one whose natural logarithm has the normal distribution. The calculations in Table 7 below indicate that the cumulative value of the lognormal evaluated at its mean is a function only of sigma. Incidently, sigma is not the standard deviation of the lognormal variate, but rather the standard deviation of its natural logarithm. If sigma is estimated to be one (a rather typical value in actual studies), Table 7 shows that only 69% of the maintenance activities will be completed by the target time, if the target is chosen to be the mean.

If, in a manner similar to what was done for the Weibull reliability case, we were to specify two desirable events, such as "we desire that the maintenance action be completed within 75

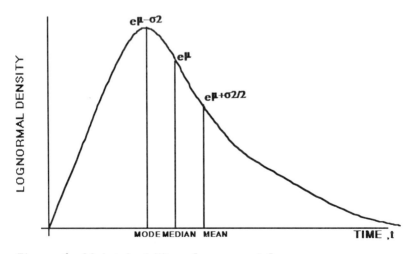

Figure 6. Maintainability : Lognormal Case

Table 7.   Lognormal Properties

$$f(t) = 1/[st(2p)^{1/2}]exp[-1/2((lnt-m)/s))^{1/2}]$$

$$F(T) = F[(lnt-m)/s]$$

$$F(e^{m+s2/2})=F[(ln(e^{m+s2/2})-m)/s]$$
$$= F(s/2)$$

If $s = 1$, $F(1/2) = 69\%$

minutes with 95% certainty" and "we desire that the maintenance action be completed in 10 minutes with 25% certainty", then the two parameters of the lognormal, mu and sigma, which completely specify the distribution, may be estimated either graphically or analytically.   Figure 7 illustrates the graphical results.

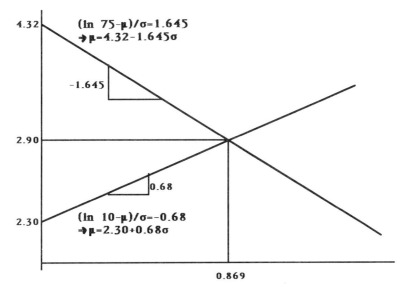

Figure 7. Suggested Reliability Goals : Lognormal Case

The point of these examples is not specifically to promote the methodology illustrated above to specify parameters of the appropriate reliability or maintainability distribution. Indeed, there are probably much better ways of doing this. The point is that those associated with design goals and those doing shop floor scheduling and predicting cycle times, where reliability and maintainability have such vital roles, not use means in their specifications.

Summary

This chapter has addressed only a few of the mathematical and statistical concerns associated with design and goal setting. The main conclusions of the work presented are: (1) Don't specify a goal without having the means to measure quantitites associated with the goal and the means to statistically test for departures from the target value, (2) First Time Yield (FTY) is a meaningful shop-floor metric, (3) although FTY can be improved by adding inspection and test stations, the improvement may be a costly one as a result of effects on capacity, throughput and cycle time, (4) PPM or DPM measures are stongly influenced by parts or operation counts and (5) reliability and maintainability goals based on the mean are extremely deceptive. A few suggestions for handling these issues have been made. It is hoped that others will recognize the problems identified here and build on the appropriate measures, tools and methodologies which will solve them.

References

Ishikawa, K., (1985) <u>What Is Total Quality Control? The Japanese Way</u>, Prentice-Hall, Englewood Cliffs, NJ.

Keats, J.B., Date, S.C. and Kim, D.Y.(1988), "Development of a Statistical Process Control Procedure and a Knowledge-Based System for Use with Attributes Data", Final Report, Digital Equipment Corporation, Phoenix, AZ and Marlboro, MA.

Lucas, J.M.,(1985) "Counted Data CUSUM's", <u>Technometrics</u>, 27,2, 129-144.

10
ANALYZING LOCATION AND DISPERSION EFFECTS FROM DESIGNED EXPERIMENTS: SOME EXAMPLES

Douglas C. Montgomery
Department of Industrial and Management Systems
Engineering
Arizona State University
Tempe, AZ 85287-5906

Abstract

Experiment design methods have traditionally dealt with determining which variables impact the mean of the process under investigation. Variables that impact the mean are referred to as location effects. In many practical situations, dispersion effects, or variables that influence process variability, are also of interest. This article surveys techniques that can be used to identify location and dispersion effects when analyzing data from designed experiments. The results of the recommended methodology are contrasted with Taguchi's parameter design and signal-to-noise ratio analysis. The results indicate that the proposed methods have better discrimination and generally require less experimentation than Taguchi's methods. The methodology is applied to industrial data in three illustrative situations.

Introduction

Statistically designed experiments are widely used in product or process development to identify which variables

157

effect the performance of the system.   For a general introduction, see Montgomery [1984, 1988].   Traditionally, experimenters have dealt with identifying factors that influence the mean of the process or system under study.   Such factors are referred to as location effects in the present paper. In many experiments, it is necessary to determine which variables impact the variability of the process.   These factors are referred to as dispersion effects.   Ideally, one could identify a set of variables that influence the mean and another set of variables that influence the dispersion in the process.   In this way, the experimenter could tune the system by using the location effects to drive the mean to an appropriate target value and then use the dispersion effects to reduce process variability as much as possible around this target.

Figure 1 illustrates this situation graphically.   In this hypothetical example, there are three process variables, A, B, and C.   The normal distributions shown at each of the eight corners of the cube represent the distribution of process measurements that would be observed at each of the treatment combinations.   Notice that as factor A moves from the low level to the high level the process mean moves off the target value

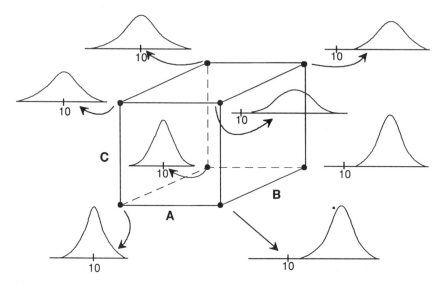

Figure 1. Location and Dispersion Effects in a Three-Variable Process

of 10. Thus, factor A is a location effect. When factor C moves from the low level to the high level the variability in the process increases. Consequently, factor C is a dispersion effect. Factor B is apparently inactive. This means that it can be adjusted to any desired level without effecting either the process average or process variability. Often, in practice, B would be run at a level to optimize cost.

Models and Methods

We assume that the experimental response y is potentially influenced by a set of variables $X_1, X_2, , X_k$. The relationship between y and these X_i is

$$y = f(X_1, X_2, X_k) + E \qquad (1)$$

where E is an error or noise term. A more convenient parameterization of (1) is

$$y = f(\underline{X}_L, \underline{X}_D, \underline{X}_N) + E \qquad (2)$$

where $\underline{X}_L$ is the subset of the X_i that influence the mean of the process (location effects), $\underline{X}_D$ is the subset of the X_i that influence the variability of the process (dispersion effects), and $\underline{X}_N$ are the variables in X_i that have negligible effects. For the example in Figure 5, A belongs in $\underline{X}_L$, C belongs in $\underline{X}_D$, and B belongs in $\underline{X}_N$.

Most conventional treatments of experiment design focus on location effects; that is, finding the variables that belong to $\underline{X}_L$. However, work on isolating dispersion effects can be traced back to Bartlett and Kendall [1946]. More recent research has been reported by Nair and Pregibon [1988], although most of their results apply to replicated experiments. We deal with the case where only a simple replicate of the experiment is available. We show by a series of examples that very simple and efficient techniques can be used to identify potential dispersion effects once the location effects in the experiment have been identified and adequately modeled.

Example 1:   The Injection Molding Problem

This example involves the use of a fractional factorial design to investigate the problem of parts shrinkage in an injection molding machine.   Parts shrinkage in injection molding is critical, as it adversely affects the final assembly of the components.   The objective of the experimenters was to reduce parts shrinkage as much as possible.   The top two panels of Table 1 show the process variables considered in this

Table 1.   Example 1. Injection Molding

VARIABLES			
	Name	Lo Level	Hi Level
A	temp	-	+
B	screw spd.	-	+
C	hold time	-	+
D	cycle time	-	+
E	moisture	-	+

RESPONSE	
Name	Units
Shrinkage	%(10)

RESPONSES		
Run Ord	Shrinkage %(10)	Std Ord
5	6	1
11	10	2
9	32	3
3	60	4
6	4	5
10	15	6
4	26	7
12	60	8
14	8	9
1	12	10
13	34	11
8	60	12
7	16	13
16	5	14
2	37	15
15	52	16

experiment and the response variable. Since each variable can be conveniently run at two levels, the design used by the experimenters is a 2^{5-1} fractional factorial design, with defining relation I = ABCDE. In this design, each main effect is aliased with a single 4- factor interaction, and each 2 factor interaction is alias with a single 3-factor interaction. The observed shrinkages (X10) from the experiment are shown in the bottom panel of Table 1.

Figure 2 presents a normal probability plot of the factor effects estimate for the injection molding experiment. From examination of this plot it is clear that factors A, B, and the AB interaction are strong contributors to parts shrinkage. Table 2 presents the analysis of variance for the selected model involving the two main effects A and B and the AB interaction. Table 3 contains the actual observed value of the response, the predicted values, and the residuals. These residuals are plotted on a normal probability scale in Figure 3. Notice that all of the residuals lie approximately along a straight line, and that there are no indications of outliers or other unusual residuals.

The effects of the factors A, B, and AB estimated in this analysis are <u>location</u> effects. For example, the estimate of A is really an estimate of how much the average value of shrinkage changes when factor A, temperature, is adjusted from the low level to the high level. Therefore the fitted model

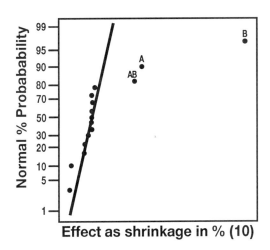

Figure 2. Effect Estimates for the Injection Molding Experiment

Table 2. ANOVA for Selected Model

SOURCE	SUM OF SQUARES	DF	MEAN SQUARE	F VALUE	PROB > F
MODEL	6410.687500	3	2136.8958333	103.086	0.0001
ERROR	248.750000	12	20.7291667		
COR TOTAL	6659.437500	15			

ROOT MSE	4.552929		R-SQUARED	0.9626	
DEP MEAN	27.312500		ADJ R-SQUARED	0.9533	

VARIABLE	DF	SUM OF SQUARES	t FOR HO EFFECT	PROB > |t|
Intercept	1		23.996	0.0001
A	1	770.062500	6.095	0.0001
B	1	5076.562500	15.649	0.0001
AB	1	564.062500	5.216	0.0002

Table 3. Predicted Values and Residuals

Std Ord	ACTUAL VALUE	PREDICTED VALUE	RESIDUAL	Run Ord
1	6.000000	8.500000	-2.500000	5
2	10.000000	10.500000	-0.500000	11
3	32.000000	32.250000	-0.250000	9
4	60.000000	58.000000	2.000000	3
5	4.000000	8.500000	-4.500000	6
6	15.000000	10.500000	4.500000	10
7	26.000000	32.250000	-6.250000	4
8	60.000000	58.000000	2.000000	12
9	8.000000	8.500000	-0.500000	14
10	12.000000	10.500000	1.500000	1
11	34.000000	32.250000	1.750000	13
12	60.000000	58.000000	2.000000	8
13	16.000000	8.500000	7.500000	7
14	5.000000	10.500000	-5.500000	16
15	37.000000	32.250000	4.750000	2
16	52.000000	58.000000	-6.000000	15

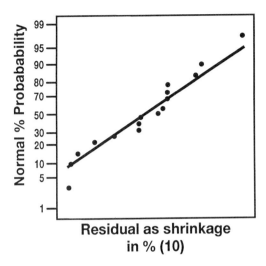

Figure 3. Normal Probability Plot of Residuals from the Injection
Molding Experiment

$$y = \hat{\beta}_0 + \hat{\beta}_1 X_1 + \hat{\beta}_2 X_2 + \hat{\beta}_{12} X_1 X_2 \tag{3}$$

where $X_1 = A_1$, $X_2 = B$, and $X_1 X_2 = AB$ can be used to predict mean parts shrinkage in the experimental region. The residuals given by

$$e = y - \hat{y} \tag{4}$$

measure the unexplained variability or noise. If there is a pattern in these residuals it could be due to dispersion effects of some process variables. One way that this could be easily checked is to plot the residuals versus each of the variables. If any variable influences variability in the residuals, then this procedure should produce a funnel-shaped plot. This graphical method of residual analysis is very simple, and is recommended in many basic experiment design books (for example, see Montgomery [1984]). It is also possible to quantify the procedure.

Table 4 displays the residuals at the low and high levels of each factor from the injection molding experiment. In this

Table 4.   Residuals at the Low and High Levels of Each Factor

-2.50	-0.50	-0.50	1.50	-2.50	-0.50	-0.25	1.75
-0.25	1.75	2.00	2.00	-0.50	1.50	2.00	2.00
-4.50	7.50	4.50	-5.50	-4.50	7.50	-6.25	4.75
-6.25	4.75	2.00	-6.00	4.50	-5.50	2.00	-6.00

−	Factor A	+		−	Factor B	+
S(A⁻) = 4.59		S(A⁺) = 3.80		S(B⁻) = 4.41		S(B⁺) = 4.01

-2.50	-0.50	4.50	7.50	-2.50	-4.50	-0.50	7.50
-0.50	1.50	-4.50	-5.50	-0.50	4.50	1.50	-5.50
-0.25	1.75	-6.25	4.75	-0.25	-6.25	1.75	4.75
2.00	2.00	2.00	-6.00	2.00	2.00	2.00	-6.00

−	Factor C	+		−	Factor D	+
S(C⁻) = 1.63		S(C⁺) = 5.70		S(D⁻) = 3.59		S(D⁺) = 4.62

-0.50	-0.50	-2.50	1.50
-0.25	2.00	2.00	1.75
-4.50	-5.50	4.50	7.50
2.00	4.75	-6.25	-6.00

−	Factor E	+
S(E⁻) = 3.40		S(E⁺) = 4.87

display, we have also calculated the standard deviation of the residuals at the low and high level of each factor. Notice that the standard deviation of the residuals at the low and high level of each factor is approximately the same, except for factor C. At the low level of factor C the standard deviation of the residuals is $S(C^-) = 1.63$, while at the high level of C the standard deviation of residuals is $S(C+) = 5.70$. This discrepancy in residual standard deviation at the two level of factor C is large enough to lead us to conclude that factor C influences the variability in parts shrinkage. This can be quantitatively demonstrated by calculating the statistic

$$\dot{F}_i = \ln\left(\frac{S(i^+)}{S(i^-)}\right)$$

(5)

for each factor considered in the experiment, as well as for each interaction of potential interest. The statistic F_i has an approximate unit normal distribution. The calculated values of this statistic for the five process variables in the injection molding are:

$$\dot F_A = 0.38, \dot F_B = -0.19, \dot F_C = 2.5, \text{ and } \dot F_E = 0.72.$$

Notice that F_C exceeds the upper 2 1/2 percentage point of the unit normal distribution, so this could be taken as evidence that factor C does influence process variability.

The method we have just illustrated in this example generalizes easily for a 2^k factorial or 2^{k-p} fractional factorial design. Each column in the design can be viewed as a collection of n/2 positive signs and n/2 negative signs. The statistic F_i compares the standard deviation of the residuals at the plus runs in this column and the standard deviation of the residuals at the minus runs in this column. In general, the equation for F_i is

$$F_i = \ln\left(\frac{\sum_{i+}(e_i - e^+)^2}{(n/2)-1}\right) - \ln\left(\frac{\sum_{i-}(e_i - e^-)^2}{(n/2)-1}\right), \qquad i = 1,2, \dots n\text{-}1 \tag{6}$$

where the subscripts i^+ and i^- imply summation only over the elements that are positive and negative, respectively, in column i of the design, and e^+ and e^- are the averages of the residuals that are associated with the positive and negative elements in column i. A version of the F_i statistic has been investigated by Box and Meyer [1986]. This statistic could be compared either to percentage points of the unit normal distribution or it could be examined via a normal probability plot. Large values of F_i indicate that factor i potentially influences process variability.

Notice that the technique we have used here effectively separates location and dispersion effects. When looking at the original responses, the standard effect estimates, based on least squares determine location effects. These locations effects are then used to build a model of the process. In the injection molding example, this model included the main effects of A, and B, and the AB interaction. The residuals from this model are a measure of unexplained error, or variability. Any pattern in this unexplained variability may well indicate that there are remaining process variables that influence the variability in the process. The statistic F_i is a useful way to examine the unexplained noise in the system for dispersion effects.

In conclusion, we have found that two process variables A and B and the AB interaction influence location in the injection problems. Since variable A (temperature) and variable B (screw speed) both have positive effects, they should be run at the low level in order to minimize average shrinkage. This conclusion is confirmed by examining the AB interaction plot. That is, the best results in terms of part shrinkage occur with AB at the lower level. Variable C (holding time) does not effect location, but it does effect dispersion in the process. The statistic F_C is positive indicating that the lowest levels of variability in parts shrinkage occur when factor C is at the low level. Factors D and E do not apparently effect either location or dispersion in the process and they may be set to any desired economical operating level.

Example 2:   The Alloy Chemistry Experiment

This is an example of a fractional factorial design used by a company making aircraft and jet turbine engine parts by investment casting. A major product development activity in this company concerns the determination of an appropriate alloy chemistry for the material. The objective usually is to achieve certain physical properties, such as high ultimate tensile strength, by varying the alloy elements of the material. In this experiment, four (4) alloying elements are considered: aluminum, titanium, chromium, and silicon. In addition, three process variables were evaluated: heat-treat, temperature, and oxygen level. The objective of the experiment was to maximize the ultimate tensile strength. The variables and response are shown in Table 5.

The engineers originally considered aluminum, titanium, chromium, and silicon content to be controllable process variables. Three process variables, heat-treat, temperature, and oxygen level, were considered to be uncontrollable variables because once the alloy went into production there

Table 5.   Example 2. An Alloy Chemistry Experiment

VARIABLES			
	Name	Lo Level	Hi Level
A	Aluminium	-	+
B	Titanium	-	+
C	Chromium	-	+
D	Silicon	-	+
E	Heat trt	-	+
F	Temp	-	+
G	o2 level	-	+

RESPONSE		
	Name	Units
R1	uts	ksi _____

was no guarantee that the same heat treating station would be used for all runs, or that the exact target temperature desired by engineering could be held and maintained by the foundry, or that the oxygen level would be exactly the same from heat to heat.   Although these process variables were uncontrollable in the actual production environment they could be controlled for purposes of a test.   Therefore, they could be used in a designed experiment.   (See Taguchi [1986]).

The   original   approach   considered   by   development engineering involved the use of a Taguchi type experiment design.   The four controllable variable were placed in an inter-array design, where each variable was run at two levels.   The design the design engineers selected was the Taguchi L_8 orthogonal array.   This design is a 2^{4-1} fractional factorial design, having eight runs.   The outer array design that was considered for the noise variables was the L_9 orthogonal array, with each of the noise variables at three levels.   Figure 4 shows the inter and outer array design that they originally considered.   Notice that this design would require 72 runs in order to obtain the desired information.

At this point, engineering personnel in charge of the study began to examine the effectiveness of the Taguchi approach. They focused on two significant issues: why are some of the

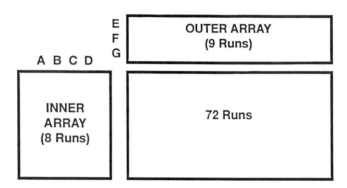

Figure 4. The Original Design the Alloy Chemistry Experiment

factors at three levels, and which interactions are likely to be important?

The reason that three factor levels are ordinarily used is to provide protection against curvature. With only two levels, the assumption is that the relationship between the response and the factor of interest is linear. Obviously, if there is curvature in the relationship between Y and a particular X, then at least three levels of X will be required to detect and model this relationship. However, when the number of variables is relatively large, it is usually a better strategy to initially concentrate on identifying which subset of factors and interactions are likely to be important rather than modeling curvature. If protection against curvature is desired, center points can be added to the two-level factorial or fractional factorial design. This will usually lead to an experiment design with far fewer runs than a three-level design. Furthermore, if the design is a fractional factorial, the two-level fractional with center parts will have a simpler alias structure than a three-level fractional.

The inter and outer array design approach recommended by Taguchi is not particularly effective in this problem. First of all, notice that it allows only interactions between the controllable variables and the uncontrollable variables to be estimated. If there are any interactions between the controllable variables they may not be discovered by this

design.   In effect, we are performing 72 runs and only
estimating a fraction of the interactions which may be of
potential interest.   A much better design strategy would be to
design the experiment so that the engineer can determine
which interactions are likely to be important without
presupposing that certain groups of interactions are negligible.
The second problem with this design is the cost.   Each part that
is used in this experiment has a foundry cost of approximately
$20,000.   Since there is some probability that the parts used in
the experiment would be unusable as production parts, and
since the Taguchi-type design would require 72 parts,
engineering and management considered this approach to be
prohibitively expensive.   Therefore, an alternate design was
used.

The design actually used in this experiment was a 2^{7-2}
fractional factorial design.   This design is of resolution IV.   All
main effects can be estimated clear of the two-factor
interactions, but a few two-factor interactions are aliased with
each other.   This design will be extremely effective at

Table 6.   Data for the Alloy Chemistry Experiment

RESPONSES			RESPONSES		
Run Ord	uts ksi	Std Ord	Run Ord	uts ksi	Std Ord
7	170.6	1	13	170.2	17
16	166	2	31	164.7	18
14	172.9	3	19	170.8	19
24	168.1	4	5	169.9	20
27	177.6	5	12	177.8	21
18	174.7	6	10	171.2	22
20	178.7	7	8	182.4	23
11	177.8	8	25	177.4	24
2	171.6	9	9	157.1	25
32	165.7	10	28	150.3	26
1	174.7	11	13	162.5	27
21	167.7	12	26	155.3	28
22	177	13	6	168.6	29
29	170.4	14	15	152.8	30
3	178.9	15	23	160	31
4	167.2	16	17	141.7	32

identifying main effects, and should give reasonable amount of information about the magnitude and extent of two-factor interaction activity. In fact, it may be possible to completely isolate all significant two-factor interactions if some of the original variables have negligible effects.

Table 6 presents the data from this fractional factorial experiment. A normal probability plot of the factor effect estimate is shown in Figure 5. This plot indicates that four main effects, A, C, E, and D are important, and that the DE interaction is also important. Table 7 presents the analysis of variance to the selected model. This analysis of variance confirms the impression of significant effects from the normal probability plot. Table 8 presents the predicted values and residuals from this experiment.

Conventional residual analysis techniques reveal no major problems with the underlying model, although there is one relatively large outlier present. Figure 6 is a normal probability plot of the residuals. The large residual is clearly

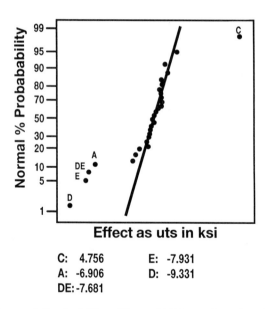

C: 4.756   E: -7.931
A: -6.906   D: -9.331
DE: -7.681

Figure 5. Normal Probability Plot of Effects for the Alloy Chemistry
          Experiment

Table 7. ANOVA for Selected Model

SOURCE	SUM OF SQUARES	DF	MEAN SQUARE	F VALUE	PROB > F
MODEL	2234.374062	5	446.8748125	25.875	0.0001
ERROR	449.033125	26	17.2705048		
COR TOTAL	2683.407187	31			

ROOT MSE	4.155780		R-SQUARED	0.8327	
DEP MEAN	168.509375		ADJ R-SQUARED	0.8005	

VARIABLE	DF	SUM OF SQUARES	t FOR HO EFFECT = 0	PROB > $|t|$
Intercept	1		229.375	0.0001
A	1	381.570312	-4.700	0.0001
C	1	180.975312	3.237	0.0033
D	1	696.577812	-6.351	0.0001
E	1	503.237812	-5.398	0.0001
DE	1	472.012812	-5.228	0.0001

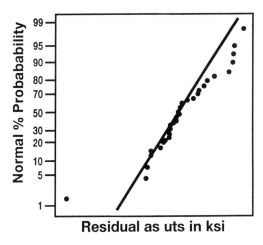

Figure 6. Normal Probability Plot of Residual

Table 8. Predicted Values and Residuals

Std Ord	ACTUAL VALUE	PREDICTED VALUE	RESIDUAL	Run Ord
1	170.600000	174.375000	-3.775000	7
2	166.000000	167.468750	-1.468750	16
3	172.900000	174.375000	-1.475000	14
4	168.100000	167.468750	0.631250	24
5	177.600000	179.131250	-1.531250	27
6	174.700000	172.225000	2.475000	18
7	178.700000	179.131250	-0.431250	20
8	177.800000	172.225000	5.575000	11
9	171.600000	172.725000	-1.125000	2
10	165.700000	165.818750	0.118750	32
11	174.700000	172.725000	1.975000	1
12	167.700000	165.818750	1.881250	21
13	177.000000	177.481250	-0.481250	22
14	170.400000	170.575000	-0.175000	29
15	178.900000	177.481250	1.418750	3
16	167.200000	170.575000	-3.375000	4
17	170.200000	174.125000	-3.925000	30
18	164.700000	167.218750	-2.518750	31
19	170.800000	174.125000	-3.325000	19
20	169.900000	167.218750	2.681250	5
21	177.800000	178.881250	-1.081250	12
22	171.200000	171.975000	-0.775000	10
23	182.400000	178.881250	3.518750	8
24	177.400000	171.975000	5.425000	25
25	157.100000	157.112500	-0.012500	9
26	150.300000	150.206250	0.093750	28
27	162.500000	157.112500	5.387500	13
28	155.300000	150.206250	5.093750	26
29	168.600000	161.868750	6.731250	6
30	152.800000	154.962500	-2.162500	15
31	160.000000	161.868750	-1.868750	23
32	141.700000	154.962500	-13.262500	17

identified on this plot. The large residual is associated with run number 32. This run has all 7 variables at the high level.

The experimenters generally suspected in advanced that this run would produce a response that was substantially lower than the rest of the runs, and that it might not be of real interest in practice. Therefore, it is relatively safe to ignore this large residual.

Since factors A and C are not involved in interactions, it is probably safe to adjust them in the direction indicated by the signs of their effect estimates. Since factor C has a positive effect and factor A has a negative effect, we would adjust C to the high level and A to the low level in order to maximize average ultimate tensile strength. To determine the appropriate levels for D and E, it is necessary to examine the DE interaction. Figure 7 presents the two-factor DE interaction plot. Notice that if factor D is at the low level, then high ultimate tensile strength is obtained, regardless of the level of factor E. This is particularly important, since factor E is the heat treating station that will be used on the factory floor and this is not easily controllable by engineering. The manufacturing personnel use the heat treating station that is available. Clearly the combination of high D and low E would also produce high ultimate tensile strength, but since we have no guarantee which heat treat station would actually be used in manufacturing, if the combination of high D and high E occurred that could lead to a heat with lower than desired tensile strength. This is an example of an interaction between a controllable process variable and a noise factor that leads to an optimal choice of the controllable factors so that the process will be robust against noise. This type of interaction will be

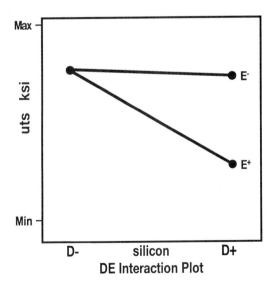

Figure 7. DE Interaction Plot

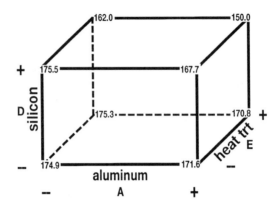

Figure 8. ADE Cube Plot

identified by the Taguchi inter and outer array approach. It will also be identified by a traditional fractional factorial design, if the experimenter arranges the design appropriately.

Figure 8 present the ADE cube plot for this experiment. Notice that high levels of ultimate tensile strength will be achieved with factor A at the low level and factor D at the low level, regardless of the level of heat treat, E. Figure 9 presents the CDE cube plot. Once again note that the combination of C at the high level and D at the low level will produce excellent ultimate tensile strength results regardless of the heat treat variable.

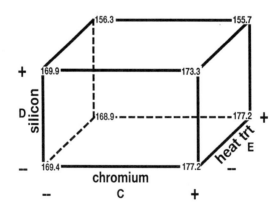

Figure 9. CDE Cube Plot

If the original Taguchi type inter and outer array experiment design had been performed, the experimenters in this problem would have likely used the signal-to-noise ratio analysis recommended by Taguchi to identify which variables influence dispersion effects in the process. Now it is well known that the Taguchi signal-to-noise ratios are not particularly effective in indentifying which variables effect process dispersion. (For further discussion, see Box [1988] and Leon et. al. [1987]). Calculation of the F statistic used in the previous example can be effective in spotting dispersion effects. Figure 10 presents a normal probability plot of the F_i statistic, applied to the residuals from the alloy chemistry experiment. Notice that there is some indication that factors D and E influence process variability. If factor D is run at the low level, this will result in the lowest levels of process dispersion attainable. Notice that factor D and E also interact, with respect to dispersion effects. Since factor E is, in effect, an

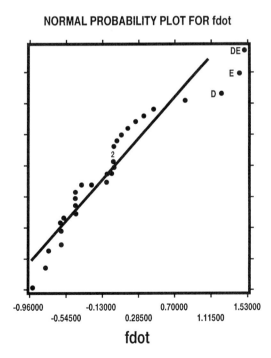

Figure 10. Normal Probability Plot for the F Statistic

uncontrollable variable, the best result that can be achieved is to set factor D to the low level. This is also consistent with obtaining high ultimate tensile strength.

Example 3: The Spring Free Height Experiment

This example was originally reported by Pignatiello and Ramberg [1985], who took the data from another source. In this problem, the original experimenters were to discover which variables affected the free height of a spring used in an automobile suspension system. Specifically, the objective was to determine which factors influenced mean spring height and which factors influenced variability in spring height in the final assembly. Five variables were considered in the original experiment design. These factors are high heat temperature (B), heating time (C), transfer time (D), hold down time (E), and quench oil temperature (O). The first four of these factors were considered to be controllable variables, and the quench oil temperature was considered to be uncontrollable. The experiment design and factor settings used by the original experimenters is show in Table 9. Notice that an L_8 orthogonal

Table 9. Example 3. Spring Free Height

Levels of Factors Studied in the Experiment				
			Levels	
Letter	Factor		Low	High
B	High heat temperature (°F)		1840	1880
C	Heating time (seconds)		25	23
D	Transfer time (seconds)		12	10
E	Hold down time (seconds)		2	3
O	Quench oil temperature (°F)		130-150	150-170

The L_8 Orthogonal Array Used and the Corresponding Free Height Data

1	2	3	4	5	6	7						
B	C	BC	D	BD	CD	E		O⁻			O⁺	
-	-	+	-	+	+	-	7.78	7.78	7.81	7.50	7.25	7.12
+	-	-	-	-	+	+	8.15	8.18	7.88	7.88	7.88	7.44
-	+	-	-	+	-	+	7.50	7.56	7.50	7.50	7.56	7.50
+	+	+	-	-	-	-	7.59	7.56	7.75	7.63	7.75	7.56
-	-	+	+	-	-	+	7.94	8.00	7.88	7.32	7.44	7.44
+	-	-	+	+	-	-	7.69	8.09	8.06	7.56	7.69	7.62
-	+	-	+	-	+	-	7.56	7.62	7.44	7.18	7.18	7.25
+	+	+	+	+	+	+	7.56	7.81	7.69	7.81	7.50	7.59

array design was used for the controllable factors, with the oil quench temperature being treated as an uncontrolled factor and placed in an outer array arrangement. The design used by the original experimenters for the controllable factors is a 1/2 fraction of the 2^4.

Table 10 presents the summary statistics obtained from this experiment design, when factor O (quench oil temperature) is treated as a controllable variable. Notice that when the design is arranged in this format it is a 16 run, two-level, fractional factorial design involving five variables. The alias structure for this design is shown in Table 11, along with the factor effect estimates for each alias pair. Notice that this design is a resolution IV experiment; that is, two factor interactions are aliased with each other. Now, an optimum one-half fraction of a two-level experiment in five variables will resolution V; that is, no two-factor interactions would be aliased with each other. This is one of the major disadvantages of the Taguchi inter and outer array approach. While it does guarantee that certain types of interactions can be estimated, there is no guarantee that the final design is of the highest possible resolution, given the number of runs that the experimenter has employed. Figure 11 presents a half-normal

Table 10. Predicted Values and Residuals

B	C	D	O	E	$\bar{y}$	s^2
			Summary Statistics for the Data			
		When O is Treated as a Controllable Factor				
-	-	-	-	-	7.79	0.0003
+	-	-	-	+	8.07	0.0273
-	+	-	-	+	7.52	0.0012
+	+	-	-	-	7.63	0.0104
-	-	+	-	+	7.94	0.0036
+	-	+	-	-	7.95	0.0496
-	+	+	-	-	7.54	0.0084
+	+	+	-	+	7.69	0.0156
-	-	-	+	-	7.29	0.0373
+	-	-	+	+	7.73	0.0645
-	+	-	+	+	7.52	0.0012
+	+	-	+	-	7.65	0.0092
-	-	+	+	+	7.40	0.0048
+	-	+	+	-	7.62	0.0042
-	+	+	+	-	7.20	0.0016
+	+	+	+	+	7.63	0.0254

Table 11. Aliases and Effect Estimates for the Spring Free Height Experiment

	COEFFICIENT	EFFECT	SUM OF SQUARES
	7.635625	7.635625	
B + CDE	0.110625	0.221250	0.195806
C + BDE	-0.088125	-0.176250	0.124256
BC + DE	-0.008125	-0.016250	0.001056
D + BCE	-0.014375	-0.028750	0.003306
BD + CE	-0.009375	-0.018750	0.001406
CD + BE	-0.018125	-0.036250	0.005256
E + BCD	0.051875	0.103750	0.043056
O + BCDEO	-0.130625	-0.261250	0.273006
BO + CDEO	0.041875	0.083750	0.028056
CO + BDEO	0.083125	0.166250	0.110556
BCO + DEO	-0.004375	-0.008750	0.000306
DO + BCEO	-0.028125	-0.056250	0.012656
BDO + CEO	0.019375	0.038750	0.006006
CDE + BEO	-0.024375	-0.048750	0.009506
EO + BCDE	0.013125	0.026250	0.002756

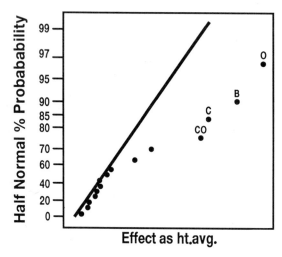

Figure 11. Half Normal Probability Plot of Factor Effects

Table 12. ANOVA for Selected Model

SOURCE	SUM OF SQUARES	DF	MEAN SQUARE	F VALUE	PROB > F
MODEL	0.703625	4	0.1759062	17.068	0.0001
ERROR	0.113369	11	0.0103062		
COR TOTAL	0.816994	15			
ROOT MSE	0.101520		R-SQUARED	0.8612	
DEP MEAN	7.635625		ADJ R-SUARED	0.8108	
C.V.	1.33%				

VARIABLE	DF	SUM OF SQUARES	t FOR HO EFFECT = 0	PROB > $|t|$
Intercept	1		300.853	0.0001
B	1	0.195806	4.359	0.0011
C	1	0.124256	-3.472	0.0052
D	1	0.273006	-5.147	0.0003
CO	1	0.110556	3.275	0.0074

plot of the factor effect estimates from this design. Notice that the only factors that affect mean spring height are O, B, C, and the CO interaction. This graphical analysis is confirmed with the analysis of variance in Table 12. The predicted values and residuals for this experiment are shown in Table 13, and Figure 12 presents a normal probability plot of the residuals. There are no indications of problems with the normality assumption, or any indication of outliers on this plot.

Figure 13 presents a graph of the residuals versus the predicted average spring heights. There is some relatively mild indication of an inwardly opening funnel on this graph, so there may be some reason to suspect that one or more process variables affect the variability in the mean spring height. Figure 14 presents the half normal plot of factor effect estimates using the log of the variance of spring height as the response variable. This graph indicates that factor B may have some effect on process dispersion. When this variable is analyzed with an analysis of variance, it does show statistical significance. Therefore, one would conclude that factor B does have some effect on the variability in spring heights. However, this variable only explains about 40% of the total variability in the system. There is no other indication that any other process variable affects the variability in spring free height.

Table 13. Predicted Values and Residuals

Std Ord	ACTUAL VALUE	PREDICTED VALUE	RESIDUAL	Run Ord
1	7.790000	7.826875	-0.036875	4
2	8.070000	8.048125	0.021875	2
3	7.520000	7.484375	0.035625	10
4	7.630000	7.705625	-0.075625	3
5	7.940000	7.826875	0.113125	1
6	7.950000	8.048125	-0.098125	6
7	7.540000	7.484375	0.055625	13
8	7.690000	7.705625	-0.015625	8
9	7.290000	7.399375	-0.109375	15
10	7.730000	7.620625	0.109375	9
11	7.520000	7.389375	0.130625	7
12	7.650000	7.610625	0.039375	12
13	7.400000	7.399375	0.000625	11
14	7.620000	7.620625	-0.000625	16
15	7.200000	7.389375	-0.189375	5
16	7.630000	7.610625	0.019375	14

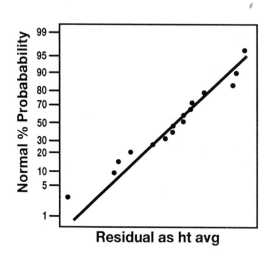

Figure 12. Normal Probability Plot of Residuals

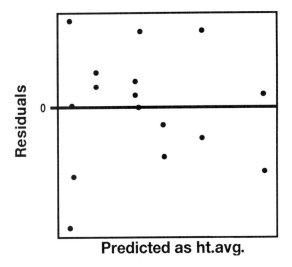

Figure 13. Residuals Versus Predicted Height

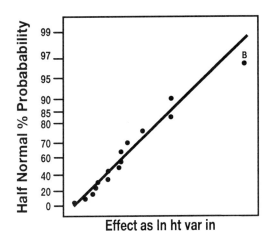

Figure 14. Half Normal Plot of Factor Effects Using ln S^2 as the
Response Variable

We may contrast this with the results obtained in the original analysis. In the original experiment, which was analyzed by using Taguchi signal-to-noise ratios and pooling all nonsignificant effects to use as an error term, the authors concluded that factors B, C, E, and O, along with the BO and CO interactions influenced mean spring height. The correct analysis of this data shows that only three of these variables (B, C, and O along with the CO interaction) really have any influence on mean spring height. The pooling techniques that Taguchi advocates often use, frequently result in too many significant factors being identified. The use of the Taguchi type signal-to- noise ratio in this experiment lead the authors to conclude that several variables may effect the process variability. Candidate variables include C, D, and B, along with the CD interaction (or its alias BE), the DO interaction, and 2-three- factor interaction alias strings, either BCO + DEO, or CDO + BEO. Contrast this with the correct analysis which only factor B is identified as having any potential influence on dispersion. Once again, this is due to the ineffectiveness of Taguchi- type data analysis methods in separating and location and dispersion effects in data from designed experiments.

References

Bartlett, M. S. and D. G. Kendall (1946). "The Statistical Analysis of Variance-Heterogeneity and the Logarithmic Transformation," Journal of the Royal Statistical Society, B, Vol. 8.

Box, G. E. P. (1988), "Signal-to-Noise Ratios, Performance Criteria, and Transformations," (with discussion), Technometrics, Vol. 30.

Box, G. E. P. and R. D. Meyer (1986) "Dispersion Effects from Fractional Factorials", Technometrics, Vol. 28.

Leon, R., Shoemaker, A. C., and R. Kacker (1987), "Performance Measures Independent of Adjustment,: (with discussion), Technometrics, Vol. 29.

Montgomery, D. C. (1984), Design and Analysis of Experiments, 2nd Edition, John Wiley & Sons, New York.

Montgomery, D. C. (1988), "Experiment Design for Product and Process Development," Manufacturing Engineering, Vol. 101, No. 3.

Nair, V. N. and D. Pregibon (1988), "Analyzing Dispersion Effects from Replicated Factorial Experiments," Technometrics, Vol. 30.

Pignaetiello, J. J. and J. S. Ramberg (1985), "Discussion of "Off-Line Quality Control, Parameter Design, and the Taguchi Method," Journal of Quality Technology, Vol. 17.

11
A QUANTITATIVE ANALYSIS RELATED TO THE EFFECTS OF FLUX TYPE ON CORROSION

Mikel J. Harry*
Chief Statistician and Member of the Technical Staff
Group Operations, Government Electronics Group
Motorola, Inc.
Scottsdale, AZ 85252

Abstract

It is well recognized that SA solder flux improves the solderability of components to a printed wiring board (PWB); however, the degree to which SA solder flux imparts corrosion is not well understood. General industry opinion seems to be split; some persons associated with soldering indicate the effects of SA flux to be detrimental enough to preclude its use, yet others believe its effect to be negligible.

The primary aim of the investigation presented in this report was to shed quantitative light on the debate over the effects of SA flux within the given expected life of a unit. It is anticipated that the results of this study will provide a decision-making framework with regard to the use of SA flux.

A statistically-designed experiment was employed to determine if there was a significant difference between RA,

*Current affiliation: Six Sigma Research Institute, Motorola University, Rolling Meadows, Illinois

RMA, and SA solder flux with respect to corrosion. The experiment was designed to allow several different types of statistical analyses to be performed; e.g., analysis-of-variance, chi-square, and k sample median test.

The various statistical analyses indicated that there was no significant difference between the three fluxes (RA, RMA, and SA). In addition, the two variable interactions were also statistically insignificant. This means that the variables are independent of each other; i.e., there are no synergistic effects. The effects of flux type was very close to being statistically significant; i.e., it appeared to be at the threshold of being an important corrosion determinant from a purely statistical view.

In terms of practical significance, the effects of flux only accounted for approximately four percent (4%) of the variability in the response variable (corrosion). Component type accounted for approximately three percent (3%) of the variation in corrosion and component height accounted for less than one percent (1%). Interestingly, all of the second order interactions explained only seven percent (7%) of the observed variation. Eighty-two percent (82%) of the observed variation in corrosion could not be explained by the variables included in the experiment; i.e., the "vital few" variables were not present in the experiment.

As a result of the aforementioned conclusions, it is reasonable to assert that even if the effect of flux type was "pushed" over the "statistical threshold," it is doubtful that the proportion of variation attributable to this effect would be of practical concern.

If the environmental testing had been more severe, then it is highly possible that the results would have provided sufficient quantitative evidence to support the notion that SA flux is a significant determinant of corrosion as compared to the RA and RMA flux types. In other words, the results and subsequent conclusions are valid only to the extent that the equivalent life of the test was in line with the basic reliability requirements of the customer.

A follow-on repeated measures, multifactor experiment should be conducted to establish the relative degree of corrosion for each of the three fluxes (RMA, RA, and SA) as a function of environmental test duration. The major purpose of this particular experiment should be the determination of the effects of environmental testing conditions; e.g., the activation mechanism for corrosion.

Since the data seems to be suspended between statistically significant differences and no statistically significant differences, the major thrust of the follow-on experiment should be to "push" the data to a decision point which is out of the "gray area."

Foreword

This research report has been divided into three (3) basic components. The first component presents a general discussion of the experiment design, statistical procedures, measurement methodology, and environmental testing procedures used during the planning and execution phases of the investigation. The second portion of the report discusses the statistical outcomes of the experiment and presents a detailed interpretation of the data. The third segment is dedicated to the conclusions and recommendations resulting from the various data analyses and subsequent interpretation of the statistical outcomes.

The narrative portion of the report discusses the experimental outcomes in a conceptual context; however, it is recognized that, even in this mode, many of the discussions and arguments may require a prerequisite understanding of experiment design and theoretical statistics. In the event the reader should require a generalized understanding of the experimental procedures, the abstract should be helpful.

In many cases, the over-generalization of experimental results represents an injustice to the reader as well as to the management who must often make critical decisions on the basis of such reports. When such over- generalization occurs, the result is often calculated in terms of unnecessary expenditures of organizational resources or, even worse, the inadvertent jeopardy of human life.

Based on the latter argument, the reader is again reminded that this particular report has not sacrificed analytical and scientific precision for the sake of reading ease.

Introduction

The experimental proceedings and analytical outcomes presented in this research report were founded on several basic questions related to the use of SA solder flux and its associated effect(s) in relation to corrosion.

Specifically, those questions were:

_Is SA solder flux significantly more corrosive than RA and/or RMA flux?

_Is component type (transistor, capacitor) associated with observable corrosion?

_Does component height play a major role with respect to corrosion?

_Does the orientation of a component significantly influence the degree to which corrosion occurs?

It is well recognized that SA solder flux generally improves the solderability of components to a printed wiring board (PWB); however, the degree to which SA solder flux imparts or facilitates corrosion is not well understood. General industry opinion seems to be split; i.e., some persons associated with soldering processes indicate the effects of SA flux to be

FACTOR LABEL	HYPOTHESIS DESIGNATOR	FACTOR DESCRIPTION	VALUE LABEL	VALUE DESCRIPTION
A	α	FLUX TYPE	1	RMA
			2	RA
			3	SA
B	*	ORIENTATION	1	0 DEGREES
			2	90 DEGREES
C	β	HEIGHT	1	FLUSH
			2	RAISED
D	γ	COMPONENT	1	LARGE TRANS.
			2	SMALL TRANS.
			3	CAPACITOR

*Suppressed during the ANOVA procedure
Note: Response variable (Y) = Corrosion

Figure 1a. Variable and Value Labels of the Experimental Factors

detrimental enough to preclude its use, yet others believe its effect to be negligible.

The primary purpose of the investigation was to shed quantitative light on the debate over the effects of SA flux within the expected life of a unit. It is anticipated that the results of this study will provide a decision- making framework with regard to the use of SA solder flux.

Figure 1a presents the variable and value labels associated with the principal independent factors of concern during this investigation. It should be pointed out that the variable/value descriptions and related value labels are applicable throughout this report; however, in some instances, variable label assignments change depending upon the analytical situation. In such cases, the reader will be directed to clarifying notation.

In order to test the significance of the variables (experimental factors) displayed in Figure 1a, statistical hypotheses were constructed. The hypotheses represent the mathematical translation of the previously stated research questions into statistical terms which are suitable for quantitative examination and testing. In essence, hypotheses are statements related to the parameters of a given probability distribution; e.g., the mean and/or variance. When stated in the null form, the general meaning is associated with the distribution of chance. This particular hypothesis is most often referred to as the "null hypothesis" and designated as "Ho." In direct contrast to the null hypothesis is the alternate hypothesis (Ha). In this form, such hypotheses are generally associated with distributions other than chance and, as such, are said to be "statistically significantly different" from the chance distribution. In other words, the hypotheses represent all possible outcomes which could result from a particular experiment.

Next, data are gathered to describe the population(s) under consideration and then statistically analyzed. Based on the statistical outcomes, each hypothesis is then accepted or rejected within certain probability limits of decision error; i.e., the results provide quantitative evidence to support the acceptance or rejection of the various hypotheses with known degrees of decision risk and confidence. Figure 1b displays the hypotheses tested in this investigation for those instances where the data were treated at the interval level.

Experimental findings do not yield narrative answers to the research questions. What the various data analyses do yield is collections of numbers within certain probability limits. The extent to which these numbers take on applied meaning is in direct proportion to the understanding of the application for which they were intended. In short, there are no magic equations that provide yes/no or black/white answers of a practical concern; only quantitative evidence to support the acceptance or rejection of the statistical hypotheses.

In order to make decisions on the basis of numbers, much gray area must be traversed. The mental integration of data and

H_0^1: $\alpha_1 = \alpha_2 = \alpha_3 = 0$

H_1^1: AT LEAST ONE $\alpha_i \neq 0$

H_0^2: $\beta_1 = \beta_2 = 0$

H_1^2: $\beta_1 \neq \beta_2$; WHERE $\beta_j \neq 0$

H_0^3: $\gamma_1 = \gamma_2 = \gamma_3 = 0$

H_1^3: AT LEAST ONE $\gamma_K \neq 0$

H_0^4: $(\alpha\beta)_{11} = (\alpha\beta)_{12} = \ldots = (\alpha\beta)_{32} = 0$

H_1^4: AT LEAST ONE $(\alpha\beta)_{ij} \neq 0$

H_0^5: $(\alpha\gamma)_{11} = (\alpha\gamma)_{12} = \ldots = (\alpha\gamma)_{33} = 0$

H_1^5: AT LEAST ONE $(\alpha\gamma)_{ik} \neq 0$

H_0^6: $(\beta\gamma)_{11} = (\beta\gamma)_{12} = \ldots = (\beta\gamma)_{23} = 0$

H_1^6: AT LEAST ONE $(\beta\gamma)_{jk} \neq 0$

H_0^7: $(\alpha\beta\gamma)_{111} = (\alpha\beta\gamma)_{112} = \ldots = (\alpha\beta\gamma)_{323} = 0$

H_1^7: AT LEAST ONE $(\alpha\beta\gamma)_{ijk} \neq 0$

NOTE It was determined that component orientation (FACTOR B) would be suppressed during the analysis-of-variance (ANOVA) procedure, thereby reducing the basic design to a 3x2x3 configuration.

Figure 1b.   Hypotheses of Experimental Concern: Three Variable Model

circumstances must prevail to the extent that consensus is achieved as to the true meaning of the data. In essence, the results of the experiment presented in this report should be viewed as a tool to guide the decision-making process rather than a vehicle for achieving concrete proofs.

Caution must be used so as not to apply the findings presented in this study to circumstances, situations, or product beyond the specific scope of the investigation. This is not to say that the data have no implications beyond the context of the experiment, but rather to say that the data are applicable, within the limits of sampling error, only to the PWBs, components, and solder fluxes defined in the report. Therefore, any generalization(s) made beyond the specific conditions of the study should be carefully scrutinized as to their inherent validity.

Procedures

To adequately answer the research questions and quantitatively test the hypotheses related to this particular study, it was necessary to execute two critical steps. First the experimental factors (variables) had to be structured such that a maximum amount of numerical information could be extracted from the basic data structure (design). Second, appropriate methods, techniques, and procedures for analyzing the experimental data had to be identified and employed to efficiently interpret the data in a quantitative and qualitative sense. This section of the report discusses the procedures related to the execution of the experiment.

Experiment Design

Full Factorial Model. The purpose of an experiment design is to structure the variables of interest for experimental manipulation. Only when the variables have been properly configured can the desired information be mathematically extracted from the data structure. In turn, the resultant quantitative information can be used to ascertain the relative contribution of the various experimental effects or "treatments."

In the case of this particular experiment, the variables of experimental concern were first configured into a 3x2x2x3 full factorial design (fixed effects model). This particular type of design allows each individual experimental factor to be evaluated independent of all other factors. In addition to the evaluation of the individual effects of the primary experimental factors, the full factorial experiment allows the interactive effects (synergistic effects) to be evaluated independent of all other effects. More specifically, the full factorial configuration structures the experimental factors in such a manner that all of the possible experimental effects can be independently examined for statistical significance; i.e., all possible treatment combinations are tested. It should be noted that, for the purposes of the analysis-of-variance (ANOVA) procedure, the 3x2x2x3 design was modified (collapsed) so as to conform to a 3x2x3 configuration. This was done to increase the rigor of the ANOVA procedure. Figure 2a illustrates the design matrix associated with the 3x2x2x3 full factorial experiment design and Figure 2b displays the collapsed configuration (3x2x3).

Hierarchical Model. In an effort to further explore the data, the basic 3x2x2x3 configuration was also modified to conform to a hierarchical configuration. The primary purpose for such an a posteriori restructuring of the experimental variables was to estimate the primary effect of each variable in a nonparametric sense. In addition to this feature of the data analysis procedure, the hierarchical model provided an adequate means to compensate for almost any experimental and/or analytical limitation which might have arisen during the conduct of the experiment; e.g., improperly prepared samples, lost data, measurement scale difficulties, and skewness. Figure 3 illustrates the basic concept of the hierarchical design configuration as used during the conduct of the investigation.

In addition to the previously mentioned features, the hierarchical configuration allowed for specified variables to be controlled (blocked) without directly contaminating the primary response. In other words, the experiment design allowed flux type to be evaluated with respect to the primary response (corrosion) free of any particular block effect(s).

By controlling the number of stages and levels of the hierarchical design structure, via the systematic stratification of certain experimental factors, the effects of various conditions could be evaluated with a high degree of analytical precision. For example, the basic design could be artificially truncated such that multiple observations would be available for one-way nonparametric comparisons and n-way cross-tablulation.

It should be pointed out that the hierarchical configuration does not allow for the two and three variable interactions to be

		A1		A2		A3	
		B1	B2	B1	B2	B1	B2
D1	C1						
	C2						
D2	C1						
	C2						
D3	C1						
	C2						

Figure 2a. Full Factorial Experiment Design: 3x2x2x3 Configuration

	A1		A2		A3	
	B1	B2	B1	B2	B1	B2
C1						
C2						

Figure 2b. Full Factorial Experiment Design: 3x3x2 Configuration

studied; however, in the case of this experiment, statistically significant interaction among the factors (second and third order) was not anticipated. With this a priori understanding, it was reasoned that the analysis of variance (ANOVA) procedure was robust enough to confirm or deny the existence of

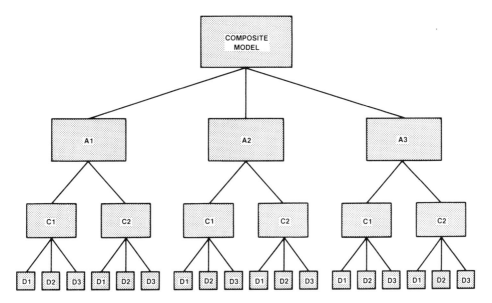

Figure 3. Hierarchical Design Model: Three Variable Configuration

statistically significant interactions in the event the assumption
of independence proved erroneous in light of nonparametric
data.

Measurement   Criteria

Prior to a discussion of the specific statistical procedures
employed to analyze the experimental data, a brief overview of
the measurement scale used to establish the relative degree of
corrosion would necessarily be in order.

Since corrosion was of primary concern in this experiment, it
was of paramount importance to establish a measurement scale
which would adequately quantify the response characteristic;
e.g., corrosion. To achieve this aim, a "Likert scale of measure"
was defined. In essence, a Likert measurement scale is a five-

point qualitative scale often used to quantify seemingly discrete attributes.

In the case of the response variable associated with this particular experiment, it would initially appear that the measurement of this variable is of a binary nature; that is to say, corrosion is either there or it is not. When corrosion is viewed in this manner, the measure (binary) is at the nominal level and, as such, does not provide enough numerical resolution to establish the relative "goodness" of an observation determined not to be corroded or the inherent "badness" of an observation determined to be corroded.

Inherent to this particular limitation is the problem of sensitivity; e.g., a binary measurement scale is not sensitive to slight variations in the universe central tendency or variance. With the aid of a Likert measurement scale, the response (corrosion) was successfully transformed from a discrete variable at the nominal level to a pseudo-continuous variable measurable at the ordinal and/or interval level. In this form, the scale was capable of measuring (detecting) the inherent "goodness or badness" of any given observation.

Another advantage of the Likert measurement scale is related to the scale's manipulative flexibility; namely, collapsing the scale into broader categories. The principal aim of exercising this feature is related to data reduction. By forcing the scale from the interval level of measure to the nominal level, the phenomenon under investigation (corrosion) could be more closely studied by altering analytical perspective.

The flexibility of the Likert measurement scale expands analytical capability which, in turn, ultimately determines the degree to which practical understanding arises. Figure 4 displays the measurement scale used to increase the sensitivity of the experiment.

Since the basic Likert measurement scale may be treated as categories, resultant data may be readily transformed to create artificial categories. In turn, this allows further analysis of a given situation or phenomenon. In the case of this investigation, such transformations were necessary to better understand the

NOTE: All data presented in this study was derived on the basis of the experimental PWB s. In addition, the observations were made by a single person from the Quality Assurance department. This particular individual was known to be associated with the project and was subsequently trained in the use of the Likert measurement scale used in this investigation.

Figure 4. Likert Measurement Scale for Corrosion

effect of the experimental factors with respect to corrosion. For example, in several cases there was not enough resolution to perform an adequate chi-square analysis due to small cell frequencies (less than 5); consequently, it was necessary to pool the Likert values of 1 and 2 into a separate category (new scale value = 0). In addition, the values 3, 4 and 5 were likewise pooled (new scale value = 1). In essence, this had the effect of forcing the original scale from the ordinal/interval level of measure to the nominal level.

The principal advantage to this type of transformation was the creation of two discrete categories ("no corrosion" and "corrosion"). This particular transformation increased the cell frequencies which, in turn, allowed for a more precise chi-square analysis. Although such mathematical transformation may initially appear to be inconsistent with the latter arguments concerning measurement scale resolution, the chi-square analyses were intended to extend the overall scope of the research. Given this purpose, such a transformation was considered to be reasonable and desirable. In addition to the enhanced analytical scope, the overall sampling/testing costs were minimized at the onset of the experiment via the a priori planning of such transformations.

One of the limitations of the five-point Likert scale is related to the underlying distribution of the response characteristic. For example, if it was determined that the optimum case associated with some "HYPOTHETICAL SITUATION" should be located at the midpoint of the scale (e.g., a value of 3), then one could reasonably expect a normally distributed population (assuming the Likert scale at the interval level of measure). This may be true for a sample of n observations randomly selected from a fixed universe at a specific point in time; but, engineering experience gained from similar experiments has shown that such an assumption often proves to be the exception rather than the rule (due to the effects of environmental testing). Consequently, provisions must be made to force the scale to other levels of measure for analysis with statistical procedures of a nonparametric nature.

In the experiment presented herein, the central tendency of the experimental data was primarily a function of the activation mechanism (environmental testing). For example, if the inherent corrosiveness of a particular flux was mildly activated after 10 days of exposure to the activation mechanism, it would be reasonable to hypothesize that the central tendency of the sample data would be at the lower end of the five-point Likert scale; e.g., no observable corrosion. From a purely statistical point of view, this would tend to suggest an underlying exponential or log-normal distribution; e.g., most of the observations would tend to cluster about the first position on the measurement scale. However, if the stress exposure were increased, 15 days, the central tendency of the sample data might likely be at the third interval on the Likert scale. In turn, this would tend to suggest a normal distribution.

Given this logic, it is possible that there may be no statistically significant difference between the three flux types for the given amount of applied stress as defined by the test requirement. However, if the test were extended, then it is possible that statistically significant differences would be observed between two or more of the flux types.

In addition to the duration of the environmental test, the nature of the underlying distribution would also depend on test severity. As a result of this line of reasoning, it was determined that the underlying distribution could not be reasonably predicted in advance of the experiment due to a general lack of a priori engineering knowledge related to the effects of the activation mechanism on the primary response variable (corrosion).

The data collected during this study were truncated and rounded prior to being analyzed. This was necessary to handle various features of the data analyses in a more efficient manner; e.g., truncation and rounding eliminated the fractional values resulting from observations which were of a multiple nature, and therein, required averaging in order to maintain a balanced design. This activity did not impact the validity of the data.

Analytical Procedures

Before presenting the analytical outcomes of the study, the statistical procedures used to analyze the experimental data are discussed. This discussion will aid in understanding the conceptual nature of the associated statistics so that insight into the data and engineering conclusions may be optimized.

Analysis of Variance. One of the primary goals of any analytical procedure is to derive quantitative information which may be used to some theoretical or practical end. To accomplish the latter aim, ANOVA was used as one of the vehicles for driving inferences back to the general population.

The major function of the ANOVA procedure is to quantitatively ascertain the portion of variability in the observed response which can be attributed to the main and interactive effects of the experimental variables as well as that portion due to random variation alone. More specifically, the purpose of ANOVA is to divide the total observed variation in the experiment into component parts; e.g., the independent contribution made by each individual variable, variable interaction, and the amount due to all other variables not included in the experiment.

After dividing the observed experimental variation into heterogeneous components, each individual part (with the exception of the total and residual) is developed into a value known as the mean square ratio (MSR) or F- value. This value is then contrasted against a theoretical value known as the "theoretical criterion" or tabulated F. The theoretical criterion represents the maximum MSR which could be expected as a function of random chance occurrence (sampling error) for a specified sample size and Type I error (alpha) probability and, in some cases, Type II error (beta) probability, delta sigma, and variance.

If the F-value derived on the basis of the experimental data were to reach or exceed the theoretical F value, then there would be sufficient quantitative evidence to support the hypothesis that the experimental effect being considered could not be explained by random variation alone; consequently, it would be concluded that the effect was due to the experimental treatment.

One of the underlying assumptions of this particular statistical procedure is that the distribution of the response variable follows a normal density function. ANOVA is rather robust to moderate violations of this assumption and, as a consequence, can provide valid information in the event that the distribution of the response measure is not normal to a certain extent..

The general model for a 3x2x3 full factorial experiment using the ANOVA procedure may be defined as follows:
$$y = m + a_i + t_j + (ab)_{ij} + (at)_{ik} + (bt)_{jk} + (abt)_{ijk} + e_{ijkl}$$
where

$$i = 1,2,...,a$$
$$j = 1,2,...,b$$
$$k = 1,2,...,c$$
$$l = 1,2,....,n$$

K-Sample Median Test. Since the nature of the underlying distribution of the response variable (corrosion) was unknown (in relation to the Likert measurement scale) prior to the execution of the experiment, it was necessary to plan for the use of nonparametric statistics. In other words, in the event the response distribution varied markedly from normality, it would have been necessary to analyze the experimental data with a statistic which was not dependent upon a normal distribution.

As evidenced by the histogram presented in Figure 5, the observed sampling distribution from one portion of the study

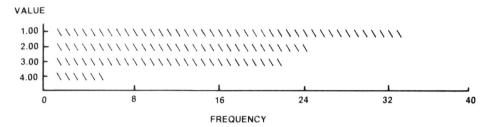

Figure 5. Truncated Inspection Data; Histogram: Pooled

did depart rather markedly from the normal density function; consequently, nonparametric statistical procedures were used to assist in the evaluation.

Since in many cases the underlying distribution of a response measure is unknown, or if known, departs markedly from normality, the application of a nonparametric statistic can prove to be of immense assistance. Under these circumstances, the arithmetic average (mean) does not provide an adequate index of central tendency; consequently, the median is used.

In general, the k-sample median test is used in cases where it is necessary to determine whether two or more populations have the same median or whether the median of the populations varies significantly from a theoretical construct (median). When this statistical procedure is used, the k-medians are pooled, and the median for the total distribution is calculated or, if testing against a theoretical construct, a hypothetical median is established. The number of observations greater than, less than, and equal to the median are counted for each sample. The frequencies are then tabulated into a contingency table and tested for statistical significance via the chi-square statistic. The hypothesis tested is that the median is the same for the k populations.

If the chi-square analysis indicates a statistically significant difference in the number of cases above/below the median in relation to the classification variable, then the null hypothesis would be rejected and the researcher would conclude that the k medians are not homogeneous. Given the latter condition, the observed difference(s) in the sample are of such a magnitude that there is an unacceptably high probability that the difference(s) could be attributable to random variation rather than any true difference in the populations under investigation.

Chi-Square Analysis. In many experiments it is often desirable to compare a set of observed frequencies with a set of theoretical frequencies. In the case of this study, it was of experimental interest to determine if the response variable was independent of the experimental (independent) variables. The question was whether or not the experimental variables (flux type, height, and orientation) were independent of the response variable (corrosion), or whether it could be argued that an association existed between any given independent variable and the dependent variable.

To make such a determination, it was necessary to treat the data at the nominal level by arranging it in the form of contingency tables, each of which was composed of R rows and C columns. Next, the data were analyzed via the chi-square statistic. The chi-square distribution is used in much the same manner as that of the normal or the F-distribution. The null hypothesis of no statistically significant difference is assumed and then tested.

To execute the test, the chi-square statistic was computed and then compared to the theoretical criterion chi- square value. If the observed chi-square value was equal to or greater than the tabulated chi-square value, Ha was accepted; i.e., there was sufficient quantitative evidence to support the notion that the experimental factor under investigation was dependent on the response variable (corrosion).

In addition to the chi-square analyses, various measures of association (correlation) were used to more rigorously examine the correlation between the independent variables and the response. These measures were inclusive of Cramer's-V and the contingency coefficient. In essence, such measures of association establish how strongly two variables are related to each other. In other words, they are measures of how well knowledge of a given characteristic's value enables one to predict the value of another characteristic or to estimate to what extent predictive uncertainty can be reduced by knowing the value of some other characteristic.

It is far beyond the scope of this report to narrate a description of the previously mentioned statistics. If a deeper understanding of the theoretical aspects of the statistics used in this study is desired, the attached references should prove useful.

Test Procedures. The testing procedures employed during this investigation involved the use of actual production boards. Two PWBs were exposed to each of the three flux types (for a total of six boards). There were eight capacitors, six TO18 transistors, and two TO5 transistors per board. Each of the devices was electrically "blown" to create open devices prior to environmental testing.

After initial resistance measurements were taken, the boards were subjected to the moisture resistance test specified in MlL-STD-202F, Method 106E. This test consisted of 10 days of temperature cycling between 25^0C and 65^0C at 90 to 98 percent relative humidity. At the end of each day, the temperature was reduced to -10^0 for 3 hours before being brought back to 25^0C. Although the vibration test is specified under this particular procedure, it was not employed since vibration was generally not believed to be a primary activation mechanism with respect to corrosion.

Corrosion measurements were made after the 10-day moisture resistance test by a quality assurance inspector. The data were then subjected to a series of parametric and nonparametric statistical analyses to test the contribution of each variable to corrosion.

The data presented in this study are valid only to the extent of the environmental testing performed. To some extent, the relative amount of environmental testing controls the degree to which corrosion occurs; e.g., it is the primary activation mechanism. Consequently, it is reasonable to assume that as environmental cycling increases, the effect of the independent variables on the response variable also increases (assuming a causal relationship).

In essence, the amount of environmental testing determines the equivalent life of the samples which, in turn, determines the validity of the test. If the test is in line with the reliability requirements of the customer, then the engineering conclusions drawn on the basis of the statistical analyses would be valid. If, on the other hand, the activation mechanism did not drive the samples to the requisite threshold, then the inherent validity of the conclusions would be applicable only to the equivalent life (as determined by the duration and severity of the test).

From an analytical perspective, the aforementioned argument is of paramount importance. For example, if the testing resolution were too low (as a result of too little environmental testing), the likelihood of decision errors would be high; e.g., the true contribution of each potential experimental effect would be unknown due to the fact that the samples would not have been exposed long enough to the activation mechanism, thereby not allowing the full effect of the causal factor(s) to surface. Thus, the equivalent effect would not be achieved according to the reliability specifications; consequently, acceptance of the null hypothesis (according to statistical convention) would almost be a certainty without knowing the true contribution of the experimental factors. On the other hand, if too many thermal cycles were used, the likelihood of committing a decision error would be unacceptably high. The true effects of the experimental treatments would be masked by the activation mechanism (environmental testing) in that the data might display statistically significant differences, but only after a point of environmental cycling far beyond the expected life of the product. In this situation, the residual sum of squares (error) would be low which, in turn, would drive high MSR values for the experimental factors. Consequently, the alternate hypotheses would be rejected in many of the cases (assuming that the environmental conditions of the test are truly activation mechanisms with respect to the independent variables).

In either of the two situations, the activation mechanism itself would be one of the controlling factors in relation to analytical resolution. Given this line of reasoning, the reader should remain cognizant of the potential implications of environmental cycling on the statistical outcomes and conclusions presented in this study.

Results

The analytic outcomes of this experiment have been partitioned into two levels. Level I presents the overall statistical results of the experiment: the effects of solder flux type. Level II presents the stratified or controlled outcomes: the outcomes related only to large transistors, a specific orientation, etc. No attempt has been made to present a detailed narrative of the specific outcomes related to the Level II analyses; however, the overall results and important findings are discussed at the end of this section.

Since the bottom line of any statistical analysis of an inferential nature involves the acceptance or rejection of scientific hypotheses, it is necessary to understand the concepts underlying statistical decision-making and reasoning.

An intrinsic component of the inferential decision-making process is the likelihood of decision error. That is, the probability of stating that a statistically significant difference exists in the universe (based on a sample) when, in reality, no such difference is present in the population. Such an error is typically referred to as a "Type I" or alpha decision error. In relation to this particular experiment, the alpha probability was established at 0.05 (5 percent).

In addition to the probability of a Type I decision error (alpha risk) is the probability of failing to detect some change as a function of the treatment. This particular type of decision error is referred to as a "Type II" or "beta" error. That is, the probability of stating there is no statistically significant difference in the universe (based on the sample data) when, in reality, a significant difference does exist. Both Type I and II risks are a measure of the likelihood of biased random sampling.

Given that there is some probability of making a decision error in almost every analytical situation (based on sample data), the major aim of any given statistical analysis of an inferential nature should be the optimization of data quality/quantity and the simultaneous minimization of estimate error.

To accomplish the latter aim, statistical procedures are used to analyze sample data which has been randomly drawn from the defined universe. The typical outcome of such an analysis is a quantitative value or number. The value or number is then contrasted against a theoretical value known as the "theoretical criterion" or "tabulated value." The theoretical criterion represents the maximum value which could be expected as a function of random chance occurrence (sampling error) for a specified sample size, variance, delta, and decision error probabilities. If the value derived on the basis of the experimental data were to reach or exceed the theoretical value, there would be sufficient quantitative evidence to support the hypothesis that the experimental effect being considered could not be explained by random variation alone; consequently, it would be concluded that the effect was due to the experimental treatment.

The outcomes of a statistical analysis serve as the primary vehicle for driving an inference back to the general population. In many cases, the purpose of such a process is to enhance the accuracy of decision making with regard to some type of action taken on one or more universe parameters; e.g., changing the setting of a machine or introducing a new material to a product in order to decrease variability. In the case of this report, the resultant action of the decision process is related to the use of SA solder flux in lieu of RA flux.

The aforementioned explanations are intended to serve as conceptual coat hangers only. It is far beyond the scope and intent of this report to present the statistical and mathematical explanations necessary for developing an in-depth understanding of the topics. If further understanding is desired, the reader is directed to the textual sources provided at the end of this report.

Level I Results

Frequency Analysis. The results of the various frequency analyses applicable to this investigation are presented in Figures 6 through 13. The histograms help to better understand the relative direction of effect for each of the three flux types in the subsequent statistical treatments of the data.

VALUE	FREQUENCY	VALID PERCENT	CUM PERCENT	PERCENT
1.00	33	27.5	39.3	39.3
2.00	24	20.0	28.6	67.9
3.00	22	18.3	26.2	94.0
4.00	5	4.2	6.0	100.0
9.00	36	30.0	MISSING	
TOTAL	120	100.0	100.0	

Figure 6. Composite Model; Truncated Inspection Data; Frequency Distribution: Pooled

The pooled data for the composite model (Figures 6 and 7) tend to support the notion that flux type is not a significant factor in relation to corrosion. The distribution is skewed right which indicates that the trend is not toward corrosion. That is, the vast majority of observations do not fall in the categories which are associated with the physical presence of corrosion (categories 4 and 5 on the Likert scale). It should be pointed out that the central tendency of the data in Figures 6 and 7, to some extent, depends on the duration and severity of the environmental test which the samples were subjected to.

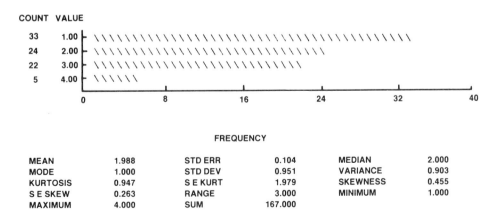

MEAN	1.988	STD ERR	0.104	MEDIAN	2.000
MODE	1.000	STD DEV	0.951	VARIANCE	0.903
KURTOSIS	0.947	S E KURT	1.979	SKEWNESS	0.455
S E SKEW	0.263	RANGE	3.000	MINIMUM	1.000
MAXIMUM	4.000	SUM	167.000		

Figure 7. Composite Model; Truncated Inspection Data; Histogram: Pooled

VALUE	FREQUENCY	VALID PERCENT	CUM PERCENT	PERCENT
1.00	13	32.5	46.4	46.4
2.00	6	15.0	21.4	67.9
3.00	6	15.0	21.4	89.3
4.00	3	7.5	10.7	100.0
9.00	12	30.0	MISSING	
TOTAL	40	100.0	100.0	

Figure 8. RMA Flux; Truncated Inspection Data; Frequency
Distribution: RMA Flux

It is evident from a cursory analysis of the histograms and
summary information presented in Figures 8 through 13 that
there is a general tendency for the RA and RMA fluxes to "push"
toward the corrosion portion of the Likert scale at a slower rate
than the SA flux (i.e., the RA and RMA flux distributions are
skewed right and the SA flux distribution is skewed left). Since
all of the samples were subjected to the same environmental
conditions and experimental treatments, it would seem that the
difference in central tendencies and skewing would be due to
flux type; however, there is no practical means to determine if
the difference is statistically significant on the basis of the
histograms.

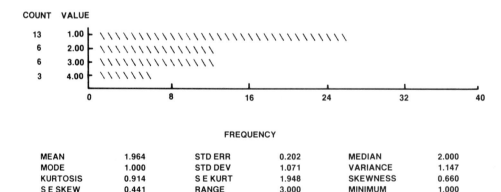

MEAN	1.964	STD ERR	0.202	MEDIAN	2.000
MODE	1.000	STD DEV	1.071	VARIANCE	1.147
KURTOSIS	0.914	S E KURT	1.948	SKEWNESS	0.660
S E SKEW	0.441	RANGE	3.000	MINIMUM	1.000
MAXIMUM	4.000	SUM	55.000		

Figure 9. RMA Flux; Truncated Inspection Data; Histogram: Pooled

VALUE	FREQUENCY	VALID PERCENT	CUM PERCENT	PERCENT
1.00	13	32.5	46.4	46.4
2.00	10	25.0	35.7	82.1
3.00	4	10.0	14.3	96.4
4.00	1	2.5	3.6	100.0
9.00	12	30.0	MISSING	
TOTAL	40	100.0	100.0	

Figure 10. RA Flux; Truncated Inspection Data; Frequency Distribution: RA Flux

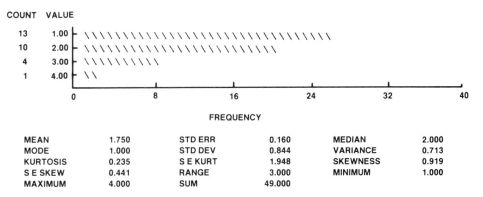

MEAN	1.750	STD ERR	0.160	MEDIAN	2.000
MODE	1.000	STD DEV	0.844	VARIANCE	0.713
KURTOSIS	0.235	S E KURT	1.948	SKEWNESS	0.919
S E SKEW	0.441	RANGE	3.000	MINIMUM	1.000
MAXIMUM	4.000	SUM	49.000		

Figure 11. RA Flux; Truncated Inspection Data; Histogram: RA Flux

VALUE	FREQUENCY	VALID PERCENT	CUM PERCENT	PERCENT
1.00	7	17.5	25.0	25.0
2.00	8	20.0	28.6	53.6
3.00	12	30.0	42.9	96.4
4.00	1	2.5	3.6	100.0
9.00	12	30.0	MISSING	
TOTAL	40	100.0	100.0	

Figure 12. SA Flux; Truncated Inspection Data; Frequency Distribution: SA Flux

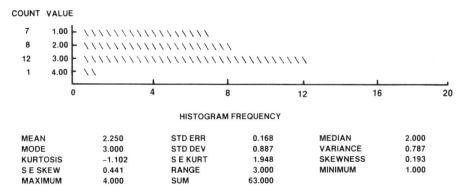

Figure 13. SA Flux; Truncated Inspection Data; Histogram: SA Flux

Analysis of Variance. The major purpose of ANOVA is to divide the variation observed in the experiment into component parts; namely, the parts attributable to the experimental treatments and the component due to error. If the resultant MSR reaches or exceeds the theoretical criterion (tabulated MSR value) for a given number of degrees of freedom, it would be concluded that the effect under consideration did not result by random chance (random sampling error).

Figure 14 displays the outcomes of the ANOVA procedure related to the investigation presented in this report.

SOURCE OF VARIATION	SUM OF SQUARES	DF	MEAN SQUARE	F	SIGNIF OF F
MAIN EFFECTS	6.423	5	1.285	1.173	0.332
FLUX	3.714	2	1.857	1.696	0.191
COMP	2.411	2	1.205	1.101	0.339
HEIGHT	0.298	1	0.298	0.272	0.604
2-WAY INTERACTIONS	6.310	8	0.789	0.720	0.673
FLUX-COMP	2.661	4	0.665	0.608	0.659
FLUX-HEIGHT	1.238	2	0.619	0.565	0.571
COMP-HEIGHT	2.411	2	1.205	1.101	0.339
3-WAY INTERACTIONS	3.054	4	0.763	0.697	0.596
FLUX-COMP-HEIGHT	3.054	4	0.763	0.697	0.596
EXPLAINED	15.786	17	0.929	0.848	0.634
RESIDUAL	72.250	66	1.095	–	–
TOTAL	88.036	83	1.061	–	–

Figure 14. Analysis of Variance Table: Rounded Inspection Data

RUN*	EXPERIMENTAL FACTOR			
	FLUX	HEIGHT	COMPONENT	
1	1	1	1	2.50
2	2	1	1	1.50
3	3	1	1	2.00
4	1	2	1	1.00
5	2	2	1	1.50
6	3	2	1	2.50
7	1	1	2	2.00
8	2	1	2	1.25
9	3	1	2	2.00
10	1	2	2	1.75
11	2	2	2	1.75
12	3	2	2	3.00
13	1	1	3	2.50
14	2	1	3	2.38
15	3	1	3	2.38
16	1	2	3	2.38
17	2	2	3	1.75
18	3	2	3	2.13

*The tabulated run order is presented in standard "experimental" order, not in the actual "randomized" run order

Figure 15. Rounded Inspection Data: Treatment Means

Figure 15 presents the treatment means (arithmetic averages).

The analysis presented in Figure 14 indicates that there is no statistically significant difference at the 0.05 level between the three fluxes (RA, RMA, and SA) given the appropriate degrees of freedom. In addition, the second order (two variable) interactions are also statistically insignificant. This means that the variables are independent of each other; i.e., there are no synergistic effects. However, the main effect of flux type was close to statistical significance. Based on the MSR probability of this particular effect (0.191) and its associated histograms, it would appear that the effect is at or near a threshold (e.g., 0.05). It is possible that a small increase in the activation mechanisms (conditions of the environmental test) could push the effect over the "statistical threshold."

In terms of practical significance, the effects of flux accounted for approximately 4 percent (4%) of the variability in the response variable (corrosion). Component type accounted for approximately 3 percent (3%) of the variation in corrosion and component height accounted for less than 1 percent (1%).

Interestingly, all of the second-order interactions explained only 7 percent (7%) of the observed variation. Eighty-two percent (82%) of the observed variation in corrosion could not be explained by the variables included in the experiment; i.e., the "vital few" variables were not among the experimental factors. As a result of the aforementioned conclusions, it is reasonable to assert that even if the effect of flux type was pushed over the potential threshold, it is very doubtful that the proportion of variation attributable to this effect would be of practical concern. However, it should also be noted that the percentages could not be much larger without also being statistically significant. In general, the percentages will most often be low, regardless of estimate error, when the effect is statistically insignificant.

In summary, the outcomes of the ANOVA procedure tended to indicate that flux type, component type, and component height are among the "trivial many" variables, with respect to corrosion measured on the Likert scale, as compared to the "vital few." As a result of the analysis, it was further concluded that, for the most part, corrosion is due to some other effect(s) not explored in this study. Furthermore, the conclusions resulting from the ANOVA procedure should be somewhat guarded by virtue of the arguments set forth within the TESTING PROCEDURES section of this report; e.g., the duration and severity of the activation mechanism (environmental testing) are strong determinants of the degree to which corrosion is observed. If the testing had been more severe, it is highly possible that the ANOVA results would have provided sufficient quantitative evidence to support acceptance of one or more of the alternate hypotheses. On the other hand, if the testing had been less severe, the MSR values would have been lower than reported. In other words, the ANOVA results and subsequent conclusions are valid only to the extent of the environmental testing and the inherent reliability/validity of the measurement scale used to quantify the response variable (corrosion) as well as the inherent robustness of the ANOVA procedure to moderate skewing in the data. It should also be pointed out that the information contained in this paragraph (related to environmental testing) is applicable to all other quantitative analyses presented in this study.

MEDIAN	FLUX TYPE		
(X)	RMA	RA	SA
CASES $> \chi$	17	15	22
CASES $\leqslant \chi$	11	13	6

$\chi^2 = 4.04;\ dF = 2; |p|\ \chi^2 = 0.1324;\ N = 84;\ \chi = 1.0$

Figure 16. K-Sample Median; Test Rounded Inspection Data By
Flux Type

K-Sample Median Test. Figure 16 presents the statistical outcomes for the comparison of solder flux type using the k-sample median test (nonparametric procedure). The reader is reminded that the k-sample median test uses the median and not the mean (arithmetic average); consequently, it cannot be interpreted in the same manner as the ANOVA outcomes presented in the previous section of this report.

In all cases involving the k-sample median test, the data were tested for statistical significance against a theoretical median of 1.0 (no observable change).

The overall conclusion to be drawn from the analytical outcome presented in Figure 16 is that there is no statistically significant difference between the three flux types (RMA, RA, and SA) with respect to the theoretical median (1.000). More specifically, the test indicated that, in relation to the first position on the Likert scale (1.0), the number of observations greater than and equal to the theoretical median were not statistically significantly different with respect to the three flux types in light of the alpha decision criterion (0.05) and the given degrees of freedom. It should be noted that the observed difference was close to being statistically significant; i.e., at or near the threshold previously mentioned.

A cursory analysis of the histograms associated with this portion of the investigation indicated that the overall central tendency of the SA flux to be generally located at the second position on the Likert rating scale; e.g., slight tarnishing or light color change, and the mode located at the third position (staining, spotting, or marked color change). This is in direct contrast to the RA and RMA fluxes. This would tend to support the notion that SA flux displayed trending toward the upper end of the Likert scale with greater momentum than the RA and RMA fluxes. In turn, this would initially appear to support the notion that SA flux did make somewhat of a "real world" difference in the amount of corrosion observed in the experiment as compared to the RA and RMA flux types; however, further consideration of the statistical outcomes of the k-sample median test indicated that the difference was not statistically significant.

The previous argument is not set forth to support the notion that the observed difference in the response characteristic (corrosion) is negligible or should be ignored, but rather to reinforce the position that there is an unacceptably high probability (approximately 20 percent) that the observed difference could have resulted by random chance variation; e.g., random sampling error. Since the observed chi-square probability is greater than the risk level established for the experiment (5 percent), it is imperative that the alternate hypothesis be rejected in the interests of scientific objectivity, regardless of natural intuition or experiential knowledge. This is not to say the null hypothesis (Ho) is true, but rather to indicate there was a dearth of quantitative evidence to support accepting the alternate hypothesis of significant difference (Ha).

Chi-Square Analysis. As previously indicated, the chi-square test ascertains if a given variable is independent of another variable or whether it may be argued that an association exists between the two variables. In the case of the Level I analysis, it was necessary to determine if an association existed between the various levels of corrosion and the three fluxes (Figure 17).

The conclusion drawn from the chi-square analysis presented in Figure 17 is that no statistically significant difference existed between the three flux types (RMA, RA, and SA). A cursory analysis of the row, marginal, and conditional frequencies shows that there is more activity in the SA flux category than in the RMA or RA categories.

CORROSION CATEGORY	FLUX TYPE			TOTAL
	RMA	RA	SA	
1	11	13	6	30
2	6	10	9	25
3	6	2	11	19
4	5	3	2	10
TOTAL	28	28	28	84

$\chi^2 = 11.46$; $dF = 6$; $|p| \chi^2 = 0.0751$

Figure 17. Cross-Tabulation and Chi-Square Analysis; Rounded Inspection Data by Flux Type

An analysis based solely on the row, marginal, and conditional (cell) frequencies would tend to imply that the SA flux is more corrosive than the RMA or RA fluxes; however, such an implication must be guided by the statistical significance of the data, the qualitative descriptions related to the five points on the scale of measure and the relative severity and duration of environmental testing. Strictly speaking, from a statistical point of view, the alternate hypothesis of significant difference would necessarily have to be rejected in this particular instance regardless of the previously mentioned mitigating factors or circumstances.

In order to better understand the degree of association between flux type and the various categories related to the Likert scale of measure, additional statistics were computed. Figures 18 display the outcomes of those computations.

The Lambda statistic presented in Figure 18 indicates the percentage of improvement in predicting the value of the dependent variable once the value of the independent variable is known. The inherent assumption associated with Lambda is that the best strategy for prediction is to select the category with the most cases (modal category), since this minimizes the number of wrong predictions. This particular approach allows the proportional reduction in error to be determined.

The maximum value of Lambda is 1.0 and the minimum is 0. A maximum value would indicate that there would be no error

STATISTIC	VALUE	DEPENDENT	
		CORROSION	FLUX TYPE
LAMBDA	0.14545*	0.09259	0.19643
UNCERT. COEFF.	0.05982*	0.05484	0.06579
CRAMER'S V	0.26119	NA	NA
CONTIG. COEFF.	0.34650	NA	NA

SYMMETRIC

Figure 18. Measures of Association; Rounded Inspection Data by Flux Type

in predicting the dependent variable by knowing the value of the independent variable. Conversely, a minimum value would indicate that the dependent variable cannot be predicted (to any extent) from knowledge of the independent variable; i.e., the two variables are independent of each other.

The reported Lambda value of 0.14545 indicates that there would be a high degree of prediction error if corrosion inspection results were to be predicted by knowing flux type. In other words, corrosion can not be adequately predicted by knowledge of flux type; i.e., corrosion and flux type are, for the most part, independent of each other.

The uncertainty coefficient is designed for nominal level variables and was applied as such in this investigation. Since it was determined to treat the Likert scale and the flux categories at the nominal level for purposes of cross-tabulation, the uncertainty coefficient was selected to further analyze the experimental data. It should be noted that the maximum value of the uncertainty coefficient is 1.0. If a value of 0.50 were to be achieved, then the interpretation would be that only 50 percent (50%) of the uncertainty in the dependent variable is reduced by knowledge of the independent variable.

The uncertainty value of 0.05484 presented in Figure 18 indicates that approximately 5 percent (5%) of the uncertainty in predicting corrosion level can be reduced by knowing the flux type. In other words, knowledge of the flux types used in the experiment did not help in knowing the corrosion outcomes observed in the experiment.

In essence, the latter two statistics in Figure 18 (Cramer's-V and the contingency coefficient) are measures of association; i.e., correlation. The reported values of these statistics indicates only a slight degree of correlation.

Chi-Square. Since the results of the ANOVA procedure, k-sample median test, and chi-square test indicated that statistical significance was at or near a threshold, it was determined that additional information would be required to solidify a concrete determination. Therefore, in order to further the aims of the study, the Likert measurement scale was collapsed into three categories for further statistical analysis using the chi-square statistic.

Essentially, this procedure involved reducing the R x C contingency table to a smaller table in order to decrease the resolution of the measurement scale which, in turn, provided greater analytical discrimination by increasing cell frequencies. Figure 19 displays the resultant Likert scale of measure. When a scale of measure is forced to the nominal level, as in the case of this experiment, and the subsequent categories are defined on the basis of the potential cell frequencies, statistical significance could be artificially induced, thereby distorting the true state of affairs. In the instance of this experiment, the Likert scale of measure was collapsed on the basis of the scale point descriptions, not the potential cell frequencies which may have resulted by virtue of the collapse.

The conclusion to be drawn from the chi-square analysis presented in Figure 20 is that the observed difference between the three flux types (RMA, RA, and SA) was not statistically significant at the .05 level with the appropriate degrees of freedom. A cursory analysis of the cell frequencies tends to indicate more activity in the SA flux category than the RMA or RA categories. In essence, this particular conclusion is the same conclusion drawn from the chi-square-square analysis prior to collapsing the Likert scale of measure.

Again, an analysis based solely on the cell frequencies would tend to imply that the SA flux is more corrosive than the RMA and RA fluxes; however, an implication must be mitigated by the statistical significance of the data, the qualitative descriptions related to interval points on the scale of measure,

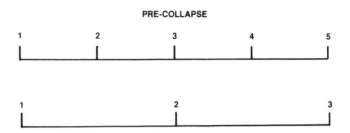

PRE-COLLAPSE

SCALE		
PRE	**POST**	**LEGEND:**
1	1	NO OBSERVABLE CHANGE
2	2	SLIGHT TARNISHING OR LIGHT COLOR CHANGE
3	2	STAINING, SPOTTING, OR MARKED COLOR CHANGE
4	3	ISOLATED DEPOSITS OR CORROSION PRODUCTS IN POWDERY OR CRYSTALLINE FORM
5	3	HEAVY CORROSION DEPOSITS OVER LARGE AREAS OR VISIBLE PITS, CRATERS OR PERFORATIONS IN METALLIC SURFACES

NOTE: The criteria for collapsing the measurement scale were related to the three basic observable characteristics; namely, (1) no observable change, (2) staining/discoloration, and (3) physical deposits.

Figure 19. Collapsed Likert Measurement Scale

CORROSION CATEGORY	FLUX TYPE			TOTAL
	RMA	**RA**	**SA**	
1	11	13	6	30
2	12	12	20	44
3	5	3	2	10
TOTAL	28	28	28	84

$\chi^2 = 6.91; dF = 4; |p| \chi^2 = 0.1408$

Figure 20. Cross-Tabulation and Chi-Square Analysis; Rounded
Inspection Data: By Flux Type (Collapsed Scale)

and the relative severity and duration of environmental testing. From a statistical point of view, Ha would necessarily have to be rejected, regardless of the mitigating factors or circumstances.

Level II Results

It is far beyond the scope and intent of this report to present a detailed narrative discussion of the Level II results; that is, the outcomes of the data analysis after stratification of the independent variables. In lieu of such a discussion, a brief discussion of the significant findings of practical concern has been presented.

In order to facilitate a more generalized analysis of the experimental data, the Likert measurement scale was collapsed a second time. In essence, the five points on the original scale were grouped into two categories;namely, "no physical deposits" designated by a 0 and physical deposits" designated by a "1." From a statistical point of view, this had the effect of inducing higher cell frequencies which, in turn, allowed the statistical significance of the resultant chi-square value(s) to be more accurately estimated. Figure 21 displays the resultant measurement scale.

Based on an analysis of the data using the collapsed Likert measurement scale (Figure 21), an alpha probability of 0.0647 for a chi-square value of 5.47712 with 2 degrees of freedom was observed when the three flux types (RMA, RA, and SA) were compared in a contingency table. Since the observed alpha probability of 0.0647 was greater than the theoretical criterion (alpha less than or equal to 0.05), the alternate hypothesis was rejected. In other words, the difference between the observed sampling distributions of the three fluxes was not large enough to be considered statistically significant; however, since the observed difference was at the threshold of statistical significance, care should be taken in concluding that the difference is not practically significant.

The board-to-board variability was statistically significant at the specified alpha criterion (0.05) with the appropriate degrees of freedom. Based on this observation, it is possible that

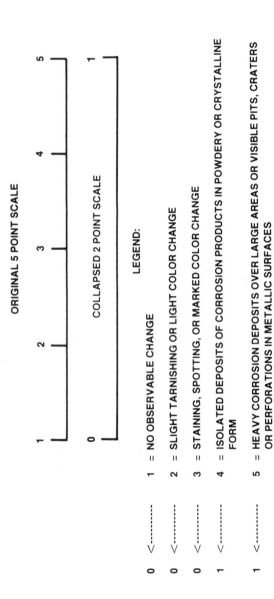

ORIGINAL 5 POINT SCALE

1   2   3   4   5

COLLAPSED 2 POINT SCALE

0   1

LEGEND:

0 <-------- 1 = NO OBSERVABLE CHANGE

0 <-------- 2 = SLIGHT TARNISHING OR LIGHT COLOR CHANGE

0 <-------- 3 = STAINING, SPOTTING, OR MARKED COLOR CHANGE

1 <-------- 4 = ISOLATED DEPOSITS OF CORROSION PRODUCTS IN POWDERY OR CRYSTALLINE FORM

1 <-------- 5 = HEAVY CORROSION DEPOSITS OVER LARGE AREAS OR VISIBLE PITS, CRATERS OR PERFORATIONS IN METALLIC SURFACES

*All data presented in this study was derived on the basis of the experimental PWB's. In addition, the observations were made by a single person from the Quality Assurance Department. This particular individual was known to be associated with the project and was subsequently trained in the use of the Likert measurement scale used in this investigation.

Figure 21. Likert Measurement Scale for Corrosion* (2 Point Version)

some variable related to the PWBs was a controlling factor, with regard to corrosion, during the experiment. This conclusion is further supported when the marginal and conditional frequencies were individually and collectively considered. It is of particular interest to note the inequitable board-to-board balancing of cell frequencies within each of the three flux categories. This tends to suggest that corrosion was controlled, to some unspecified extent, by other factor(s) not included in the experiment in that the experimental results (corrosion) were not repeatable within a treatment when the collapsed measurement scale was employed.

When component height was considered, a statistically significant difference between the three flux types was noted at the 0.05 (5 percent) level of significance given the associated degrees of freedom when the variable was controlled to the low level (flush mount). When the control variable was considered at its high level, there was a lack of evidence to support the alternate hypothesis of significant difference. In essence, this particular analysis indicates that corrosion is, to a moderate extent, dependent upon whether or not a component is raised off of the PWB or whether it is flush mounted. An analysis of the marginal and conditional frequencies indicated that the PWBs that received the RA flux treatment were less likely to show visible signs of corrosion than the PWBs treated with the other two flux types (RMA and SA). Further discussion of this effect is unnecessary since flush mounted components generally are known to be associated with "flux entrapment."

Overall, it would appear that the experimental outcomes based on an analysis of the stratified data supports the notion that the flux type is not, for the most part, a controlling factor with regard to corrosion. The effects of the activation mechanism control the statistical outcomes to some extent. If the activation mechanism was severe enough to extend the samples beyond the required life, then the effects of flux type should be considered negligible; however, if the converse is true, then the latter statement would not hold true.

Conclusions

The investigation presented in this report was initiated to answer several fundamental questions related to solder flux.

*Is SA solder flux significantly more corrosive than RA and RMA flux?

*Is component type (transistor, capacitor, etc.) associated with observable corrosion?

*Does component height play a major role with respect to corrosion?

*Does the orientation of a component significantly influence the degree to which corrosion occurs?

Since the conclusions related to each particular analysis were narrated after the presentation of the associated statistical outcomes (refer to the RESULTS section of this report), no attempt has been made to rearticulate the findings within this portion of the paper. However, the basic implications of those conclusions have been summarized below in the context of the research questions.

Question 1: Is SA solder flux significantly more corrosive than RA and RMA flux?

Answer: It is generally believed that the cleaning operation used during the experiment eliminates most of the residue which may facilitate corrosion when the data were considered as a composite (i.e., no stratification). This statement must be mitigated by the severity and duration of the activation stress (temperature, humidity, and time, respectively). If the conditions of the environmental test are in line with the basic reliability requirements of the customer, then the latter generalization will hold true. Although, the effects of flux types were close to being statistically significant, the percent of variation which this factor explained was very low.

Question 2: Is component type (transistor, capacitor, etc.) associated with observable corrosion?

Answer: Based on the experimental data, it does not appear that the effects of the components which were considered in the experiment are significantly related to corrosion; however, a general analysis of the contingency table frequencies would tend to indicate that, although not statistically significant, the large transistors tended to maintain a higher propensity to display physical signs of corrosion than the capacitors or small transistors.

Question 3: Does component height play a major role with respect to corrosion?

Answer: Based on the experimental data, it appears that the effects of component height plays a role with respect to corrosion. Although statistically significant, this factor accounted for a very small portion of the total observed corrosion. Again, this would tend to imply that the cleaning process is adequate.

Question 4: Does the orientation of a component significantly influence the degree to which corrosion occurs?

Answer: Given the qualitative and quantitative interpretations of the experimental data, it does not appear that component orientation significantly influences the degree to which corrosion occurs. However, it should be noted that there was a slightly higher propensity for those components which were oriented at a 90 degree angle to Figure physical signs of corrosion. Since the transistors included in the experiment were round, it is generally believed that the effect is due primarily to various aspects of the cleaning operation; i.e., the cleaning process is not uniform with respect to all four sides of the PWB.

Recommendations

The following recommendations are based on qualitative and quantitative evaluations of the experimental results, situation, and circumstances.

A follow-on repeated measures, multifactor experiment should be conducted to establish the relative degree of corrosion for each of the three fluxes (RMA, RA, and SA) as a function of environmental test duration. The major purpose of this particular experiment should be related to determining the effects of the environmental testing conditions; i.e., the activation mechanism. Since the data seems to be suspended between statistically significant differences and "no statistically significant differences," the major thrust of the follow-on experiment should be to push the data to a decision point that is out of the gray area.

If corrosion is primarily due to the activation mechanisms and not the experimental variables, then the resultant alpha probabilities should not change much. On the other hand, if the effects of the experimental factors increase with additional environmental testing and the effects of the test prove not to be causal, then the alpha probabilities should change radically. In either case, greater analytical resolution will be present which, in turn, will provide more information from which to make a managerial decision with regard to the production use of SA flux.

A total of four boards will be required for the follow-on experiment. These boards should be good boards from the same vendor lot. Care should be taken to make sure the boards are as identical to each other as possible. These boards should be handled as carefully as possible to avoid confounding the test results. In addition, preparation and testing of the samples should be executed as quickly as possible; i.e., the boards should not be idle for an extended period of time.

Experimental Procedure

1. Obtain four good boards from the same vendor lot. Care should be taken to ensure that the boards are as identical as possible. Rejected boards should NOT be used due to uncertainties surrounding the effects of tin/lead thickness on corrosion.

2. Label each board with sequential numbers starting with the number 1. Prepare a measurement sheet for each board.

3. Measure tin/lead thickness at 10 randomly selected points on each board prior to parts insertion. The results of this

measurement should be recorded on the measurement sheet with location and board number.

4. Insert eight small transistors in each board. Four transistors should be raised off the board and four should be mounted flush.

NOTE: The small transistor configuration from the first experiment should be used.

5. Wave solder one board using RMA flux, one board using RA flux, and two boards using SA flux.

6. Clean all boards using standard cleaning procedures.

7. Select one board prepared with SA flux and repeat the standard cleaning procedure two additional times. Immediately place this board in a sealed plastic bag to prevent environmental contamination.

8. Submit all four boards to a 10-day environmental stress test.

9. Upon completion of the environmental stress test, inspect all boards (top and bottom of joints and at tin/lead thickness measurement points). Rate each inspection point using the 5-point measurement scale and record results on measurement sheet. Three raters should perform the inspection. One of the raters should be the same person who performed inspections on the previous experiment. All observations must be derived by agreement. Observations that cannot be agreed upon should be noted by an asterisk (*).

10. Following the corrosion inspection, all boards should be resubmitted to environmental test and be reinspected; i.e., repeat steps 8 and 9 for all boards.

11. Upon completion of the second inspection, all boards should be tested for cleanliness using an Omega meter or equivalent contaminant testing procedure. The results of the cleanliness test should be recorded on the measurement sheet for each board.

12. Submit testing results for statistical evaluation.

NOTE:Preliminary results (after first environmental test and inspection) should be submitted prior to the second environmental cycle.

References

Box, G.E.P.; Hunter, W.G.; and Hunter, J.S. (1978). Statistics for Experimenters. John Wiley and Sons, Inc., New York, NY.

Daniel, C. (1976). Application of Statistics to Industrial Experimentation. John Wiley and Sons, Inc., New York, NY.

Davies, O.L. (1978). The Design and Analysis of Industrial Experiments. Longman, Inc., New York, NY.

Hunter, J.S. (1985). "Statistical Design Applied to Product Design." Journal of Quality Technology, 17, pp. 210-221.

Phadke, M.S.; Kackar, R.N.; Speeney, D.V.; and Grieco, M.J. (1983). "Off-Line Quality Control for Integrated Circuit Fabrication Using Experimental Design." The Bell System Technical Journal, 62, pp. 1273-1309.

1 2
AN EXPERIMENTAL STUDY OF THE EFFECTS OF VARIOUS PROCESS AND DESIGN CHARACTERISTICS ON SOLDER JOINT CRACKING

Mikel J. Harry*
Chief Statistician and Member of the Technical Staff
Group Operations, Government Electronics Group
Motorola, Inc.
Scottsdale, AZ 85252

Abstract

This report presents the outcomes of two factorial experiments designed to surface the principal cause-and- effect relationships underlying solder joint heel cracks related to a particular transistor. More specifically, the report presents the results of a fractional and full factorial experiment designed to explore the main and interactive effects of several process and design-related variables with the aid of analysis of variance (ANOVA) and related graphical procedures.

The first experiment was structured to isolate the vital few variables related to solder cracks. The purpose of the second experiment was to further explore statistically significant effects identified in the first experiment and any other factors believed to significantly influence the primary response

*Current affiliation: Six Sigma Research Institute, Motorola University, Rolling Meadows, Illinois

parameter (solder cracks). In addition, the second experiment was intended to serve as a validation study.

Based on the analytical outcomes of the experiments, there is evidence that rather marked variations in solder deposition characteristics and assembly technique will not unduly increase the likelihood of a solder crack if the the stress loop configuration related to the specified transistor is used. However, this does not appear to be the case when the G-lead configuration is used. In this configuration, there appears to be a significantly greater likelihood of observing solder cracks under various conditions of preloading and different amounts of solder. The latter effect appeared to be even more marked when the J-lead configuration was used.

Given this observation, an increase in the stress relief capability of the lead, as a function of lead configuration, decreases the likelihood of solder cracking regardless of the level of preloading and solder quantity to a certain extent. In other words, solder cracking is primarily due to random variation when the stress loop is present; however, if the stress loop is not present, then solder cracking depends, to a great extent, on certain other variables; e.g., preloading and solder joint quality.

Introduction

The purpose of this report is to present the outcomes of two separate experiments designed to surface the principal cause-and-effect relationships underlying solder joint heel cracks. More specifically, the report presents the results of a 2^{5-1} fractional and 2^5 full factorial experiment (subsequently collapsed into a 2^3 configuration) designed to explore the main and interactive effects of five process and design related variables with the aid of analysis of variance (ANOVA) and various graphical procedures.

The first experiment was structured to isolate the "vital few" variables related to the response. The second experiment was structured to further explore the statistically significant effects identified in the first experiment and any other factors believed to significantly influence the primary response parameter (solder cracking). In addition, the second experiment was intended to serve as a validation study.

Problem

The problem was centered around solder joint cracks developing at the heel of a certain transistor. This particular problem involved high reliability equipment designed for space and ground applications. The problem surfaced as the direct result of two field failures. The major thrust of the first investigation (2^{5-1} fractional factorial) involved determining the effect of the five variables identified in Figure 1 on solder joint integrity.

The second experiment (2^5 full factorial) was aimed at determining the effects of the experimental factors listed in Figure 2 on solder joint integrity.

Objectives

To resolve the problem of solder joint integrity (heel cracks), two statistically designed experiments were conducted: a 2^{5-1} fractional and 2^5 full factorial experiment. The research objectives associated with the two experiments have been located in Figure 3 of this report.

Factor Code	Factor Description	Experimental Level	
		Low (−)	High (+)
A	Amount of Gold on the Lead	Dip Once	Dip Three Times
B	Component Lead Preloading	None	10 Mil
C	Staking Material Thickness	20 Mil.	60 Mil.
D	Lead Configuration	"J" Lead	"G" Lead
E	Solder Joint Quality (Solder Deposition)	Minimal	Full

Figure 1. Experimental Factors Associated with the 2^{5-1} Fractional Factorial Design (First Experiment)

Factor Code	Factor Description	Experimental Level	
		Low (–)	High (+)
A	Staking Material Thickness	20 Mil	60 Mil
B	Component Lead Configuration	Stress Loop	"G" Lead
C	Component Lead Preloading	None	20 Mil
D	Conformal Coating	None	Present
E	Solder Joint Quality	Minimal	Full

Figure 2. Experimental Factors Associated with the 2^5 Full Factorial Design (Second Experiment)

The factors for both experiments were selected on the basis of their individual and joint likelihood of causation (with respect to solder cracking). Since the basic nature and scope of the problem was unknown at the time the investigation was undertaken, it was necessary to advance theories which could offer plausible explanations of the phenomenon and therein identify critical parameters for subsequent experimental verification. Figure 4 illustrates the three different lead configurations associated with factor "D" during the first experiment and factor "B" during the second experiment.

Objective Number	Objective Description
1	Determine the main and interactive effects of the experimental factors, each at two levels
2	Ascertain the proportion of variation in solder joint cracking attributable to the independent and joint effects of the experimental factors
3	Determine the optimum treatment combination of the experimental factors in relation to the response characteristic (solder joint cracking)
4	Provide an estimate of the likelihood of decision error with regard to objective 1, 2, and 3.

Figure 3. Research Objectives

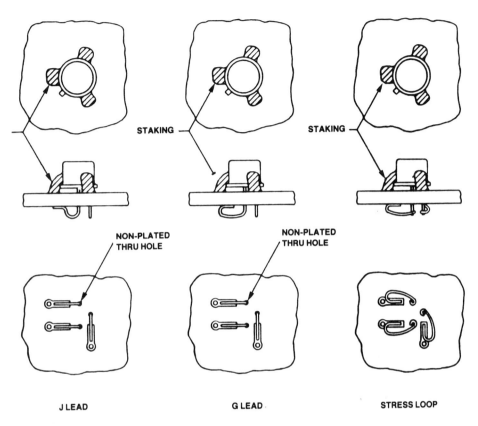

Figure 4. Illustration of Experimental Levels Related to Lead
           Configuration

Procedures

The experimental strategy used to isolate the cause-and-effect relationships consisted of running two separate factorial experiments. The first experiment involved structuring the primary independent variables into a 2^{5-1} fractional factorial experiment design. Figure 5 illustrates the design matrix associated with this particular configuration.

A fractional factorial design, in a broad sense, is essentially a carefully prescribed subset of treatment combinations drawn from a full factorial configuration - a fraction of the treatment

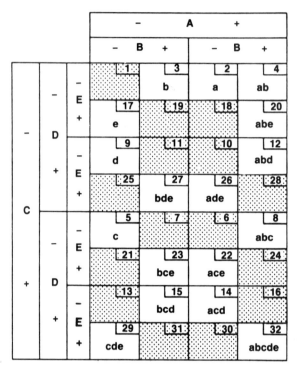

Figure 5. 2^{5-1} Fractional Factorial Design Matrix

combinations available under the full factorial counterpart. With this type of design configuration, some of the interactions are confounded with other experimental effects; i.e., the unique individual effects of the confounded experimental factors cannot be separated for independent analysis. Under this condition, the individual contribution of a given experimental effect is said to be "mixed" with some other experimental effect(s).

In those cases involving four or more experimental factors it is common practice to assume third order and higher interactions to be statistically insignificant or negligible. This was believed to be a reasonable assumption in the instance of the research presented herein.

Since a full factorial experiment involving five factors has the capability to produce 32 separate effects, of which half are third order and higher interactions, it makes more analytical sense to assess only the main effects and second- order interactions (assuming that replication or an external source of experimental error is prcsent or available). Based on the preceding logic it was determined, in the instance of the first experiment, that third-order and higher interactions were negligible (given that the primary purpose of the first experiment was to screen for leverage). As a result of these and other considerations, a 2^{5-1} fractional factorial design was selected to structure the experimental factors for testing.

The parent of the fractional factorial experiment design is the full factorial configuration. In essence, a full factorial design structures the experimental factors so that all possible experimental effects can be quantitatively examined for statistical significance; i.e., all possible treatment combinations are tested. Herein lies the principal advantage of the factorial design over the classical one-factor-at-a-time approach; i.e., the capability to structure the variables so that the interactive effects can be ascertained. Figure 6 illustrates the design matrix associated with the 2^5 full factorial experiment.

Since the first experiment (fractional factorial) surfaced several statistically significant effects, it was determined that a second experiment should be conducted to validate the aforementioned results and test additional factors for statistical significance. In the case of the second experiment, each of the 32 treatment combinations were replicated three times. This was done to control for Type Il decision errors; e.g., beta risk. The number of replicates was derived by mathematical computation involving specified values of alpha risk, beta risk, delta sigma, and variance. The variance estimate used in the computation of sample size for the second experiment was made on the basis of the first experiment. Delta sigma was established a priori by the engineering team.

In order to determine if any particular parameter maintains a relationship with a given phenomenon, it is necessary to collect and scientifically analyze numerical data. Figure 7 depicts the data collection sheet used to organize the numerical information collected during the conduct of the experiment. The primary task of any data analysis procedure, as used in

			− A		+ A	
			− B	+ B	− B	+ B
C −	D −	E −	1 (1)	3 b	2 a	4 ab
		E +	17 e	19 be	18 ae	20 abe
	D +	E −	9 d	11 bd	10 ad	12 abd
		E +	25 de	27 bde	26 ade	28 abde
C +	D −	E −	5 c	7 bc	6 ac	8 abc
		E +	21 ce	23 bce	22 ace	24 abce
	D +	E −	13 cd	15 bcd	14 acd	16 abcd
		E +	29 cde	31 acde	30 bcde	32 abcde

Figure 6. 2^5 Full Factorial Design Matrix

conjunction with a statistically designed experiment, is to extract numerical information from the design configuration. The outcomes of the extraction process should facilitate an understanding of the phenomenon under investigation. Based on the outcomes of the previously mentioned process, the investigator can make inferences and apply them to the population from which the samples were drawn.

Inherent to such an inferential process is the likelihood of decision error; e.g., the probability of stating that a significant difference exists in the universe parameter of interest (based on a sample) when, in reality, no such difference exists. Such an error is typically referred to as a Type I or "alpha" decision error. Given that there is some probability of making a decision error in almost every analytical situation (based on sample data), it becomes readily apparent that the major aim of any

LINE	EFFECT	RUN	A	B	C	D	E	(I) E	(I) B	(I) C	(II) E	(II) B	(II) C	(III) E	(III) B	(III) C	STRESS LOOP Y	STRESS LOOP N
1)	TOTAL	24	−	−	−	−	−											
2)	a	18	+	−	−	−	−											
3)	b	22	−	+	−	−	−											
4)	ab	32	+	+	−	−	−											
5)	c	4	−	−	+	−	−											
6)	ac	11	+	−	+	−	−											
7)	bc	28	−	+	+	−	−											
8)	abc	23	+	+	+	−	−											
9)	d	15	−	−	−	+	−											
10)	ad	1	+	−	−	+	−											
11)	bd	8	−	+	−	+	−											
12)	abd	27	+	+	−	+	−											
13)	cd	19	−	−	+	+	−											
14)	acd	31	+	−	+	+	−											
15)	bcd	2	−	+	+	+	−											
16)	abcd	25	+	+	+	+	−											
17)	e	30	−	−	−	−	+											
18)	ae	16	+	−	−	−	+											
19)	be	7	−	+	−	−	+											
20)	abe	29	+	+	−	−	+											
21)	ce	17	−	−	+	−	+											
22)	ace	20	+	−	+	−	+											
23)	bce	13	−	+	+	−	+											
24)	abce	6	+	+	+	−	+											
25)	de	5	−	−	−	+	+											
26)	ade	12	+	−	−	+	+											
27)	bde	10	−	+	−	+	+											
28)	abde	21	+	+	−	+	+											
29)	cde	26	−	−	+	+	+											
30)	acde	14	+	−	+	+	+											
31)	bcde	3	−	+	+	+	+											
32)	abcde	9	+	+	+	+	+											

FACTOR CODE	FACTOR DESCRIPTION	FACTOR LEVELS (−)	FACTOR LEVELS (+)
A	STAKING MATERIAL THICKNESS	20 MIL SPACING	60 MIL SPACING
B	LEAD CONFIGURATION	STRESS LOOP	6 LEAD
C	PRE-LOADING	ABSENT	20 MIL
D	CONFORMAL COATING	ABSENT	PRESENT
E	SOLDER JOINT	MINIMAL SOLDER JOINT	FULL SOLDER JOINT

STRESS LOOP

Y – STRESS LOOP PRESENT
N – STRESS LOOP ABSENT

REPLICATES

E - EMITTER
B - BASE
C - COLLECTOR

SOLDER JOINT CRACK SCALE OF MEASURE

1 – PERFECT SOLDER JOINT
2 – HEEL CRACK PRESENT
3 – CRACK FROM HEEL TO 25% OF TOTAL LENGTH
4 – CRACK FROM HEEL TO 50% OF TOTAL LENGTH
5 – CRACK FROM HEEL TO 75% OF TOTAL LENGTH
6 – CRACK FROM HEEL TO MORE THAN 75% OF TOTAL LENGTH
7 – 100% HEEL CRACK

Figure 7. Data Collection Sheet for Second Experiment

statistical analysis of an inferential nature should be the optimization of data quality/quantity and the simultaneous minimization of estimate error.

To accomplish the latter aim, analysis of variance (ANOVA) was used as the primary vehicle for conveying inferences back to the general population. Essentially, the major function of the ANOVA procedure is to quantitatively ascertain the portion of variability in the observed response which can be attributed to the main and interactive effects of the experimental variables as well as that portion due to random variation alone. More specifically, the purpose of ANOVA is to decompose the total observed variation in the experiment into component parts; e.g., the independent contribution made by each individual variable, variable interaction, and the amount due to all other uncontrolled causal variables not included in the experiment.

After decomposing the observed experimental variation into heterogeneous components, each individual part (with the exception of the total and residual) is developed into a value known as the mean square ratio (MSR) or "F" value. This value is then contrasted against a theoretical value known as the "theoretical criterion" or "tabulated F". The theoretical criterion represents the maximum MSR which could be expected as a function of random chance occurrence (sampling error) for a specified sample size and Type I error (alpha) probability. If the F-value derived on the basis of the experimental data were to reach or exceed the theoretical F-value, there would be sufficient quantitative evidence to support the hypothesis that the experimental effect being considered could not be be explained by random variation alone. Consequently, it would be concluded that the effect was due to the experimental treatment.

In addition to the probability of Type I decision error (alpha risk) is the probability of failing to detect some change as a function of the treatment. This particular type of decision error is referred to as a "Type II" or "beta" error; e.g., the probability of stating there is no statistically significant difference in the universe parameter of interest (based on the sample data) when, in reality, a significant difference does exist. As may be apparent at this point, both types of risk (Types I and II) are a measure of the likelihood of biased random sampling.It should be pointed out that the aforementioned explanations are intended to serve as conceptual coat hangers. It is far beyond the scope and intent of this report to present the statistical and mathematical explanations necessary for developing an in-depth understanding of the topics. If further understanding is desired, a bibliography of textual sources has been provided at the end of this report.

The testing procedures used during both experiments have been summarized in Figure 8. Each of the listed steps entailed many substeps; however, only the higher order steps have been included in this report.

It should be pointed out that data were collected after each iteration of step 3. This was necessary to ascertain the point at which the analytical resolution would be great enough to adequately analyze the data in light of the effects of environmental testing. For example, if the resolution were too low (as a result of too few thermal cycles), the likelihood of decision error would be excessive. That is, the true contribution of each experimental effect would be unknown due to the fact that the samples had not cycled enough to induce an effect. On the other hand, if too many thermal cycles were used, the likelihood of committing a decision error would also be

Step Number	Step Description
1	Random selection of parts
2	Sample Preparation
3	Thermal cycle and vibration
3	Thermal cycle and vibration
4	Inspect samples and record data
5	Analyze data
6	Go to step 3 if resolution is too low

Figure 8.   General Testing Procedures

unacceptably high. That is, the true effects of the experimental treatments would be masked by the failure mechanism (thermal cycling) in that all of the solder joints would display signs of physical fatigue. In either of the two possible situations, the failure mechanism itself would be the controlling factor in relation to analytical resolution; consequently, iterations of step 3 were made until it was believed that an adequate amount of analytical resolution had been achieved.

The measurement standard employed to assess the response (solder cracking) was a 5-point qualitative scale often referred to as a Likert scale of measure. This type of measurement scale is very useful in quantifying seemingly discrete attributes. In this case, it would appear that a solder crack is of a binary nature; i.e., it is either there or it is not. In essence, the Likert scale of measure allowed the response to be translated from a discrete variable at the nominal level to a pseudo-continuous variable measured at the interval level. The overall effect of this particular translation greatly enhanced the analytical resolution of the experiment and significantly decreased the overall sampling/testing costs. See Figure 7 and 8 (lower left hand corner) for an explanation of the rating criteria related to this portion of the study.

Results

The first portion of this section presents the analytical outcomes of the 2^{5-1} fractional factorial experiment. The reader should note that the results of the first experiment have been broken down by component lead: collector, emitter, and base. In this particular instance, the data were gathered on the individual leads of the transistor in order to determine if the experimental treatments maintained a uniform effect on each of the leads. It should be pointed out that the Likert measurement scale was applied to the experimental solder joints by three individuals known to be closely associated with the product, the aims of the experiment, and the measurement scale criteria.

The second portion of this section narrates the results of the second experiment (full factorial). The second experiment was conducted primarily to validate the significant findings of the first experiment and explore other potential sources of variation. As a result of this mission, the second experiment was conducted somewhat differently in terms of the rating criteria and data analysis. A 7-point rating scale was used in lieu of a 5-point scale, and the data from the individual leads were pooled into an aggregate for purposes of data manipulation and analysis. The purpose of the latter effort was to circumvent various potential problems associated with Type II decision errors.

The following narrates the experimental outcomes of the fractional and full factorial experiments, respectively. An alpha risk of 10 percent (Type I error probability) was employed as the decision criterion with regard to all of the experimental effects. Given the alpha risk, associated degrees of freedom, and the extent to which the data were valid and representative of the population, the following effects were noted as statistically significant. Furthermore, Figure 9 illustrates the experimental effects through the use of a cell mean plots related to the various treatment averages from the full factorial experiment.

		PERCENT (%) OF
EXPERIMENT 1 (FRACTIONAL FACTORIAL):		EXPLAINED
		VARIATION
o	Base lead	
	- Lead configuration	38.3%
	- Solder joint quality	7.0%
	- Staking material thickness and lead preloading interaction	7.0%
o	Emitter	
-	No effects were statistically significant at the 0.10 level; however, lead configuration was within close proximity	0.0%
o	Collector	
-	No effects were statistically significant at the 0.10 level; however, lead configuration was within close proximity	0.0%
	TOTAL	52.3%

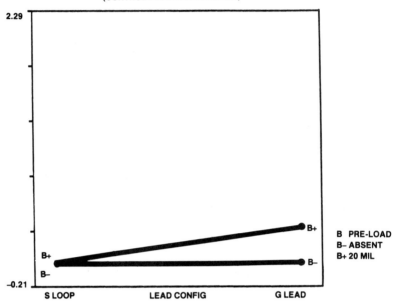

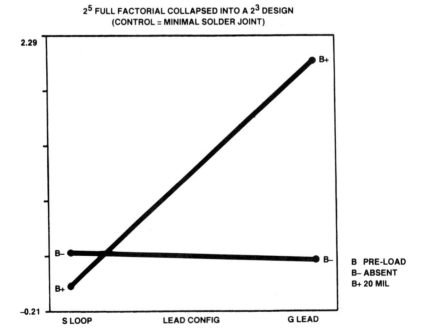

Figure 9.   Plot of Cell Means for the Collapsed 2^5 Full Factorial Experiment

EXPERIMENT 2 (FULL FACTORIAL):

* Lead configuration   4.9%
* Lead preloading   3.0%
* Solder joint quality   4.9%
* Lead configuration and preloading interaction   5.6%
* Lead configuration and solder joint quality
  interaction   2.5%
* Lead configuration, lead preloading, and
  solder joint quality   3.0%

                                   TOTAL   23.9%

NOTE:The percent of explained variation in the second experiment (full factorial) is considerably less than the first experiment (fractional factorial) primarily because the "J" lead was omitted from the second experiment. Since the stress loop lead configuration demonstrated superior performance, it is reasonable to expect less activity in the response variable and, as a result, display lower percentages; e.g., the main effect of the factor (lead configuration) and its associated interactions would be greatly diminished, thus explaining a lower proportion of the observable variation.

The following general notes are related to the results of the fractional and full factorial experiments:

Thermal cycling/vibration was the primary failure mechanism; i.e., the principal driver of the response (solder cracks). As a result, the measurement rating that any particular solder joint received was a function of the experimental treatment and the amount of environmental testing the joint had been subjected to. As a consequence of this, the percent of variation attributable to a particular treatment was, for the most part, a function of the amount of environmental testing experienced by the solder joint at that point in time. It is generally believed that if the samples had been subjected to more stress, the percent of unexplained variation would be significantly less. Due to testing costs, the experiment was terminated after enough analytical resolution was realized to achieve the research aims.

In effort to better understand the analytical outcomes of the full factorial experiment (second experiment) it was necessary to accomplish two tasks. First, the data related to the individual leads at each treatment combination was pooled into a composite value. This was necessary since it was believed the stress loop lead configurations would considerably decrease the propensity for solder joint cracks (as compared to the lead configuration used during the first experiment) at the same level of environmental testing and, as a consequence, generate less active data for statistical analysis. Second, the 25 experiment design was collapsed into a 23 configuration. The primary reason for this procedure was to increase the error component of the ANOVA. In turn, this had the effect of providing a more rigorous test of the significance of the MSR.

Implications

Prior to a presentation of the engineering implications, the reader is reminded that implications are a conceptual extrapolation of the data to the situation under investigation. It is important to understand that the extrapolation is made by interlinking the experimental results to create a more complete understanding of the data.

EXPERIMENT 1 (FRACTIONAL FACTORIAL):

Implication 1: The factors considered in the investigation do not affect the three types of leads in a uniform manner; consequently, it would appear that there are other factors impacting the response which were not included in the experiment; e.g., inherent assembly/processing differences.

Implication 2: The reported percent of variation in the response explained by the treatments is not a good index to judge the intrinsic weight of a particular variable in the total scheme. This form of reasoning maintains merit since environmental cycling was the primary failure mechanism; i.e., it is reasonable to assume that the percent of variation explained by any given variable at some point in the testing sequence is relative to that point in time only. Consequently, the proportional differences in the relative effects at the terminal point of acceptance testing should be the primary index to assess the relative weight of any particular variable. For example, if there were an absence of environmental testing, then the amount of explained variation (as a function of the experimental factors) would be extremely small and, as a result, appear inconsequential. This would also appear to hold true at the opposite end of the continuum with respect to environmental testing. Given this line of thought, it would appear that there would be some point during the environmental testing at which the percent of explained variation would be at a maximum level. Since environmental testing was not continued to the point of declining returns, it was not possible to derive the true proportion of variation explained by the experimental factors. As a consequence, the differential between proportions should be used as a basis for determining the degree of influence a particular experimental factor exerts in relation to all other such factors. Based on this form of logic, it would be reasonable to believe that the data contains· valuable information about reliability differences resulting from altered process and design conditions (treatments).

EXPERIMENT 2 (FULL FACTORIAL):

Implication 1: Based on the analytical outcomes of the experiment, there is evidence to believe that rather marked variations in solder deposition and assembly technique will not unduly increase the likelihood of a solder crack if the stress loop configuration is used. However, this does not appear to be the case when the G- lead configuration is used. Under this condition, there appears to be a significantly greater likelihood of observing solder cracks under various conditions of preloading and different amounts of solder. The latter effect seems to be even more marked when the J-lead configuration is used. Given this observation, it may be generally understood that an increase in the stress relief capability of the lead, as a function of lead configuration, will decrease the likelihood of solder cracking regardless of the level of preloading and solder quantity (within limits). In other words, solder cracking is primarily due to random variation when the stress loop is present; however, if the stress loop is not present, then solder cracking is dependent upon certain other variables; e.g., preloading and solder joint quality.

Implication 2: Since there were several solder cracks present in the stress loop configuration after only 17 thermal cycles, and given that the stress loop configuration (as an experimental level) displayed optimum performance, it is believed that the cracking was not a function of the failure mechanism or the stress loop configuration but rather some unidentifiable nonconformity in the process or material at the time of sample preparation. It should be noted that the data, as a whole, tends to support this notion. In addition, numerous authors on the subject of experiment design indicate that it is not unusual for this type of situation to arise under experimental conditions.

References

Bowker, A., and Lieberman, G. (1972). Engineering Statistics. Englewood Cliffs, N.J.: Prentice-Hall.

Burr, I. (1953). Engineering Statistics and Quality Control. New York: McGraw-Hill Book Co.

Cochran, W., and Cox, G. (1957). Experimental Designs (2nd edition). New York: John Wiley and Sons.

Daniel, C. (1976). Applications of Statistics to Industrial Experimentation. New York: John Wiley and Sons.

Dixon, W., and Massey, F. (1969). Introduction to Statistical Analysis (3rd edition). New York: McGraw-Hill Book Co.

Draper, N., and Smith, H. (1966). Applied Regression Analysis. New York: John Wiley and Sons.

Guenther, W. (1973). Concepts of Statistical Inference. New York: McGraw-Hill Book Co.

Hahn, G., and Shapiro, S. (1967). Statistical Models in Engineering. New York: John Wiley and Sons.

Lipson, C., and Sheth, N. (1973). Statistical Design and Analysis of Engineering Experiments. New York: McGraw-Hill Book Co.

Mood, A. and Graybill, F. (1963). Introduction to the Theory of Statistics (2nd edition). New York: McGraw-Hill Book Co.

Ward, J. and Jennings, E. (1973). Introduction to Linear Models. Englewood Cliffs, NJ: Prentice-Hall.

Winer, B. (1979). Statistical Principles in Experimental Design (2nd edition). New York: McGraw-Hill Book Co.

SECTION III

PROCESS ANALYSIS
AND IMPROVEMENT

1 3
A SIMULATION STUDY OF STATISTICAL PROCESS CONTROL ALGORITHMS FOR DRIFTING PROCESSES

Robert V. Baxley, Jr.
Monsanto Chemical Company
Pensacola, Florida 32575

Abstract

Time series control methodology has gained acceptance in many process industries for quality variables measured at discrete time intervals. These algorithms calculate a series of adjustments which are designed to compensate for drifting disturbances which would otherwise drive the process away from target. These controllers call for an adjustment for every sample interval, which may be too costly if the adjustments cannot be automated. At the opposite extreme are the Shewhart control charting procedures which employ a hypothesis testing approach aimed at minimizing the risk of taking action when the process is in control (Type I error). When drifts are present, however, the Shewhart procedures have a relatively high control error sigma (product variability) compared with time series controllers. This paper studies feedback control strategies which represent a compromise in the sense of requiring less frequent adjustments than time series control but providing closer control to target than Shewhart control charts.

Drifting behavior is simulated using a first order integrated moving average (IMA) model fed by standard normal shocks from a random number generator. The rate of drift away from target is varied by changing the IMA parameter, r. CUSUM and EWMA control strategies are evaluated in order to find tuning parameters which minimize the control error sigma for specified average adjustment intervals from 5 to 20 sample periods. It was found that control error sigma increases (1) with the average adjustment interval and (2) when one period of deadtime is present in the feedback loop. Both of these effects are more severe as the rate of drift increases. The EWMA control algorithm was found to have a slightly lower control error sigma than CUSUM control for any given average adjustment interval within the range studied.

Introduction

Background. The interest in minimizing product variability has been growing in recent years, sparked by the effort of the automobile industry to improve quality in response to foreign competition. Statistical Process Control (SPC) methods have been a key ingredient of this effort, as evidenced by Ford Motor Co. [1980), and these methods have propagated into the automotive supplier chain . The main reason being that supplier audits look for evidence that processes are in statistical control as defined originally by Shewhart [1931), and that they are capable of consistently meeting specifications as measured by the C_p index defined by Kane [1986] as the ratio of the specification range divided by six times the process capability sigma. This has required that even small shifts in either the process mean or variability be detected and corrected as soon as possible, and as a result the SPC literature is replete with enhancements to the conventional Shewhart control chart which will on average detect a small shift sooner. The three most important such charting enhancements are (1) the use of runs rules based on recent historical data as popularized at Western Electric Co. [1956], (2) Cumulative Sum (CUSUM) Charts first introduced in England by Page [1954] and popularized at DuPont by Lucas [1973, 1976], and (3) Exponentially Weighted Moving Average (EWMA) Charts introduced by Roberts [1959].

The performance of these charting methods has been evaluated by using the average run length (ARL), which is the average number of sample intervals from the time a level shift occurs until it is signalled by the chart. It is desired that a charting method have a large average run length when the process is in control, so that there will be few false alarms. On the other hand, it is also desirable to have a short average run length when a shift has occurred. Champ and Woodall [1986] used this average run length criterion to conclude that the CUSUM chart is superior to Shewhart charts with runs rules for detecting small level shifts, and Lucas and Sacucci [1987] have found EWMA and CUSUM charts to have approximately the same run length characteristics.

Characteristic of the ARL approach is the assumption that a manufacturing process is characterized by periods of stable operation where (common) causes of the system result in random variations about target. These periods are interrupted by special causes which result in instantaneous shifts in the process mean away from target. In many process industries the product quality data do not fit this characterization. Instead, there are forces which are always present and result in a drifting behavior. Bagshaw and Johnson [1974,1975] treated these drifts as part of the common cause variation using the coefficient of serial correlation which measures lack of independence between adjacent observations. In calculating the effects of serial correlation on the run length distribution of CUSUM charts, they found that control limits should be widened in order to avoid a decrease in the in-control average run length. Bissell [1984], on the other hand, calculated the effects on run length distributions of Shewhart and CUSUM charts when the special cause is manifested as a ramp rather than a step change. In the same manner as for step changes, he found that CUSUM charts have a shorter average run length than Shewhart charts for detecting ramps with small or moderate slopes.

Another approach for handling drifting processes, is to move away from the null hypothesis of random variation about a fixed mean toward a class of stochastic time series models called autoregressive integrated moving average (ARIMA) models of Box and Jenkins [1976]. These ARIMA models are used to characterize and forecast the drifting behavior of process disturbances when no control action is taken and to describe the dynamic relationship between a controlled and a manipulative variable. From them a feedback control algorithm can be derived which minimizes the variance of the controlled variable by making an adjustment at every sample point which exactly compensates for the forecasted disturbance. These time series control algorithms are compared with conventional PID (proportional, integral, derivative) controllers by Palmor and Shinnar [1979].

Hunter [1986] recognized a continuum of control strategies ranging from Shewhart control charts to these time series controllers. He suggested the exponentially weighted moving average (EWMA), a subclass of ARIMA disturbance models, as being particularly applicable for quality data in many process industries. Box, Jenkins, and MacGregor [1974] show that a control strategy comparing the EWMA to a set of fixed limits is optimal when the cost of making adjustments is significant compared with the cost of being off target. In doing so, they have calculated the effect of changing the EWMA limits on the average adjustment interval and the control error variance, for the case where the adjustment takes effect immediately.

Objective.   In many instances, particularly in the process industries, close control of a laboratory measured variable to the specification target is necessary to meet customer requirements, and yet it would require too much production operators' time to make an adjustment for every sample. Also, it is common that the effect of the adjustment is not seen in the next sample but is delayed until the following sample. This is called deadtime in the feedback loop and can be caused by delays in performing the laboratory measurement or by processes which take a long time to deliver material from the point of adjustment to the sample point. The objective of this paper is to use simulation methods for such processes in order to study the performance of feedback control algorithms which require less frequent adjustments than time series controllers and yet have only a slight increase in control error variability.

Process Simulation Modeling

The variation in a process quality measurement can be represented as coming from two sources: (1) outside disturbances caused by variation in raw material, uncontrolled variation in the process itself and measurement error; and (2) the effects of process adjustments made for the purpose of controlling the quality index at a target. Figure 1 is a block diagram showing how these sources of variation are related to the feedback control logic. The reasoning underlying the construction of simulation models for these two factors is presented below.

Outside Disturbances. This simulation study does not consider sudden level shifts due to special causes not considered a part of normal system variation because such causes should be identified and eliminated in order to achieve a process improvement. The intention of the feedback control action, rather, is to minimize the system variation, which in the process industries includes drifts along with measurement error. For modeling this type of system variation we choose the integrated moving average process of order (0,1,1), which is defined in Box and Jenkins [1976, p. 105] by the difference equation,

$$N_t = N_{t-1} + a_t - q \, a_{t-1}, \tag{1}$$

where q is a parameter to be estimated from process data, N_t is the effect of the disturbance and the a_t's are $NID(0, s_a^2)$. This model can often be used to describe laboratory data in the process industries because it can also be represented as follows:

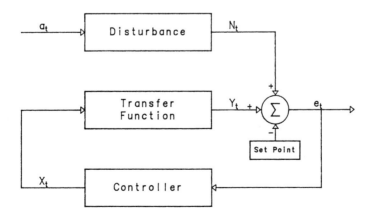

Figure 1. Block Diagram of the Feedback Control Simulation Model

$$N_t = z_t + b_t \qquad (2)$$

where the b_t's are white noise, $NID(0, s_b{}^2)$, representing sampling and measurement error, and z_t is a random walk process representing the combined effects of lurking variables which tend to drive the process away from target. The random walk is an integrated moving average $(0,1,1)$ process with $q = 0$:

$$z_t = z_{t-1} + u_t \qquad (3)$$

where the u_t's are $NID(0, s_u{}^2)$. A process with fast drifts is one where z_t, which represents the true process level at time t, can move rapidly over the range equal to the magnitude of s_b, because of a relatively large variance of the shocks, u_t. In other words, the forces driving the process away from target are large relative to the sampling and measurement errors. The integrated moving average parameter, q, which has values from zero to one, is closer to zero for such a process. In contrast, processes with slow drifts have a small variance of u_t and larger sampling and measurement errors. They can be modelled as an integrated moving average with q closer to one. Box and Jenkins [1976, p. 124] give the following equation relating q to the ratio of the variances of u_t and b_t :

$$s_u{}^2/s_b{}^2 = (1-q)/q \qquad (4)$$

Simulated data from slow, moderate, and fast drifting processes are shown on Figure 2.

Another feature of the integrated moving average model is that, as shown by Box and Jenkins [1976, p. 144], Equation 1 can be rewritten to give the following recursive formulas for updating a forecast of the process level, l periods into the future:

$$\hat{N}_t(l) = \theta \, \hat{N}_{t-1}(l) + (1-\theta) \, N_t \qquad (5)$$

and

$$\hat{N}_t(l) = \theta \, \hat{N}_{t-1}(l) + (1-\theta) \, a_t \qquad (6)$$

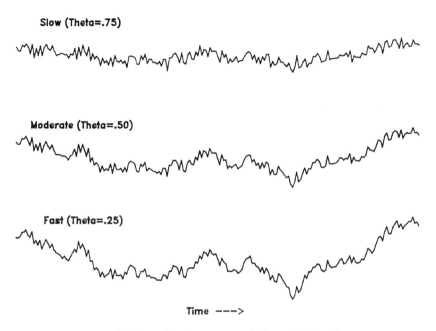

Figure 2. Plot of Slow, Moderate, and Fast Drifts Time

In this notation, t represents the time at which the forecast is being made, and 1 represents the time forecast in terms of the number of sample intervals into the future. In other words, the forecast is made at origin t for leadtime 1. The hat over N denotes an estimate of the expected value of N for any future time, conditional upon the realizations of N up to time t. Equation 5 shows that the forecast originating at any time (t) is a weighted average of the previous forecast (time t-1) and the current data point. Equation 6 shows that the expected process level is only partially impacted by each new shock, a , whose effect is reduced by the factor (1-r). As shown by Box and Jenkins [1976, p. 128] these forecasts estimate the process deviation from target without bias, and the forecast errors,

$$e_t(l) = N_{t+l} - \hat{N}_t(l)$$

have a lower variance than those for any other statistic calculated from historical data. Hence they are useful from a process control standpoint because they help determine the appropriate adjustment for returning the process to target.

Although the forecasts are the same for all future sample points (values of l), the forecast error variance increases with l. This can be seen by expressing the forecast errors as a linear combination of future shocks, again as shown in Box and Jenkins [1976, p. 128]:

$$e_t (l) = y_0 \ a_{t+1} + y_1 \ a_{t+l-1} + \ldots\ldots + y_{l-1} \ a_{t+1}, \qquad (7)$$

where $y_1 = y_2 = \ldots..y_{l-1} = (1-q)$ and $y_0 = 1.$

Because the a_t's are independent with equal variances, and the fact that for such variables x_1 to x_k the variance of any linear combination of them is as follows:

$$\text{Var} \ (b_1 \ x_1 + \ldots\ldots + b_k \ x_k \ ) = (b_1{}^2 + \ldots\ldots + b_k{}^2 \ ) \ s^2,$$

the forecast error variance is:

$$\text{Var} \ [e_t \ (l)] = [1 + (l-1) \ (1-q)^2] \ s^2 \qquad (8)$$

A forecast of simulated data from an integrated moving average process is shown on Figure 3, along with the 95% confidence limits at

$$\hat{N}_t(l) \pm 2 \ (\text{Var} \ [e_t(l)])^{1/2}$$

Finally these forecasts are a weighted average of current and historical data, where the weights are decreasing exponentially for data further back in time. This can be seen by re-expressing the current forecast of Equation 5 by making successive substitutions for the previous forecasts to obtain the following:

$$\hat{N}_t(l) = (1-\theta) \ (N_t + \theta \ N_{t-1} + \theta^2 + \ldots\ldots\ldots.)$$

from which the weights are seen to be $(1-q)$, $(1-q)q$, $(1-q)q^2$, ..etc., which sum to unity for $0 < q < 1$ because of the fact that

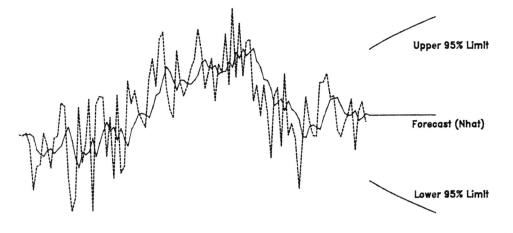

Time --->

Figure 3. Forecasting an Integrated Moving Average Process
(Theta=.75)

$$\sum_{i=0}^{\infty} x^i = 1/(1-x), \qquad 0 < x < 1$$

As shown by Figure 4, these weights are relatively heavy on recent data for fast drifts but spread further back in time for slow drifts.

The Effects of Process Adjustments. For many laboratory measurements the effect of a process adjustment is immediate in the sense that when an adjustment is made because of a particular sample result, that adjustment will have completely taken effect before the next regularly scheduled sample is obtained. This relationship is shown in Figure 5 relating in time the manipulative variable X to its effect on the process, Y. For other laboratory measurements, however, there may be delays relating to sample delivery, the time required to perform the analysis, or data movement to the person making the adjustment. Also, the adjustment may be made at an early processing stage prior to vessels with large hold-up time so that

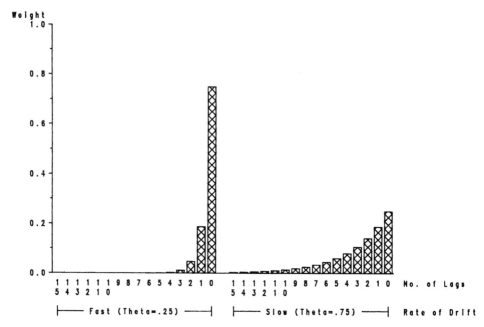

Figure 4. Forecast Weights for IMA(0,1,1) Model

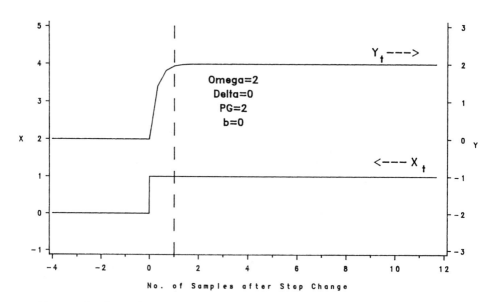

Figure 5. Step Response with No Deadtime or Carryover

the adjustment effects require some time to reach the sample point and do not take full effect immediately. These situations are better represented by Figure 6. Box and Jenkins [1976, p. 350] suggest the following difference equation for these effects of process adjustments:

$$Y_t = d\ Y_{t-1} + w\ X_{t-b-1}$$

This equation can be rewritten in the form of a transfer function as follows:

$$Y_t = \frac{\omega}{1-\delta B}\ B^b\ X_{t-1},\tag{9}$$

where B is the backshift operator such that

$$BX_t = X_{t-1}$$

and

$$B^b\ X_t = X_{t-b}$$

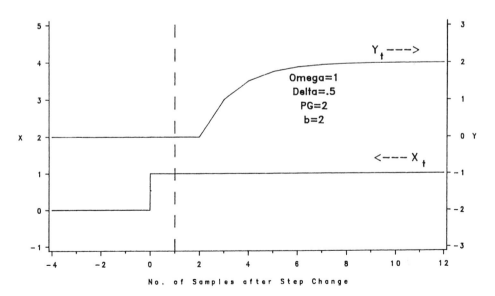

Figure 6. Step Response with Deadtime and Carryover

The transfer function is made up of two components: B , which is pure deadtime during which no effect of the adjustment is seen, and

$$\frac{\omega}{(1 - \delta B)},$$

which is the dynamic response of the process. The parameter, w, is the magnitude of the response to a unit step change in the first period following the deadtime, and d $(0 \leq d < 1)$ is the factor measuring carry- over of the exponential process response into succeeding sample periods. The process gain, PG, is the eventual effect of a unit change in the manipulative variable after the dynamic response has been completed. It is calculated as follows:

$$PG = w (1 + d + d^2 + ...), \quad 0 \leq d < 1$$

$$PG = w /(1 - d) \tag{10}$$

The right side of Equation 10 can be factored as follows:

$$PG = \omega + \delta \frac{\omega}{(1-\delta)}$$
$$= w + d\, PG$$

in order to show how the process gain is divided into first period and carryover effects. If d=0, then PG= w because no carryover effects take place. As shown by Box and Jenkins [1976, p. 359], a slight generalization of this model is needed in order to represent fractional periods of deadtime prior to initialization of the exponential response.

Alternative Control Algorithms

In order for a statistical control algorithm to be included in this study, it was necessary to have a method of calculating the process adjustment from statistics calculated from historical data. This ruled out the modified Shewhart approach which splits the control chart into six zones and employs run rules (8 in a row on one side of target, 4 of 5 outside of one standard deviation, etc.) because they take a hypo- thesis testing approach to the decision of whether or not the process mean is on target. If the null hypothesis (on target) is rejected, there is no statistic built into the control logic which provides an estimate of the new process level, and hence no adjustment can be calculated. The three types of statical algorithms which meet the above criterion are (1) Time Series control, (2) Exponentially Weighted Moving Average control, and (3) Cumulative Sum control. A discussion of each of these controllers follows.

Time Series Controllers. Time series controllers require an adjustment for every sample but provide a performance benchmark because they give the minimum control error variance as long as their underlying process models are correct. They are derived mathematically from the disturbance and transfer function models with the objective of minimizing the mean squared deviation from target of the quality index. This is accomplished by positioning the process in order to exactly compensate for the forecasted deviation from target at the time when the current adjustment will take effect, b+1 periods into the future. If this is done, the deviation from target or control error, e , is just the error from a forecast originating b+1 periods in the past.

As the level of the manipulated variable at time t (X) is placed to compensate for the forecast, the adjustment or change in the manipulated variable is calculated to compensate for the change in the forecast from the previous sample period. Therefore, the adjustment for a Time Series controller where the disturbance is $IMA(0,1,1)$ and there are no carry- over effects (w=0) is minus the inverse of the process gain times the change in the (b+1)-step-ahead forecast,

$$x_t = -(1/w) (1- q) a_t$$

as seen by substituting from Equations 6 and 10. Because the a_t 's are not observed, we must re-express the above in terms of the control (or forecast error), e_t. From Equation 7, we have $e_t = a_t$ for b=0 (P=1), and hence the control equation is

$$x_t = - (1/w)(1- q)e_t \qquad (11)$$

This is recognized as integral action by control engineers, but the magnitude of the adjustment is reduced because the disturbance model says that only (1 - q) * 100 percent of the control error will affect the future process behavior.

Using the same method for b=1,

$$e_t = a_t + (1 - q) \, a_{t-1}$$

and rearranging in terms of the backshift operator we have

$$a_t = \frac{e_t}{1 + (1 - \theta) B}$$

Therefore, the control equation for one period of deadtime is

$$x_t = -(1/w)(1 - q)e_t - (1 - q)x_{t-1} \qquad (12)$$

The first term is still the integral action, and as shown by Palmor and Shinnar (3), the second term is the deadtime compensator developed by Smith [1959]. An appendix of the article by Palmor and Shinnar gives a method for deriving Time Series control equations for the general class of disturbance and transfer function models.

The reason that time series controllers provide an important baseline for studying controllers requiring occasional adjustments is that their control error variance is known to be the (b+1)-step-ahead forecast error variance which can be calculated from Equation 8 for the IMA disturbance model.

$$\mathrm{Var}[e_t \, (b+1)] = [1 + b(1 - q)^2] \, s_a^2 \qquad (13)$$

For the case of no deadtime (b=0), the control error variance is seen from Equation 13 to be equal to variance of the random component of the disturbance process, s_a^2. The effect of deadtime is to increase the control error variance by an amount which depends on q. Figure 7 shows this relationship in terms of the control error sigma for zero, one, and two periods of deadtime. It can be seen that for slow drifts (q › 1) the deadtime causes only a small increase in the control error sigma, while for fast drifts (q › 0) there is a large such increase. For processes with fast drifts, then, it is important to

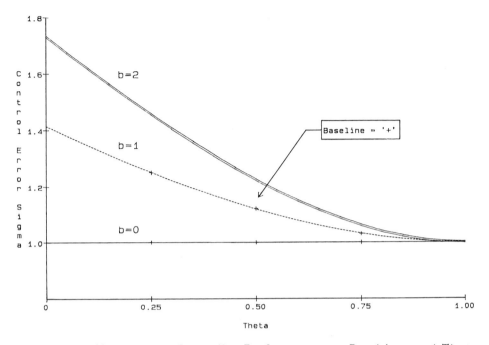

Figure 7. Time Series Controller Performance vs Deadtime and Theta
with Sigma(a) = 1.0

reduce dead- time. The six points on Figure 7 identified with the symbol, +, represent baseline controller performance levels against which the other controllers requiring less frequent adjustments will be compared via simulation.

Exponentially Weighted Moving Average Control. The idea of EWMA control originally advanced by Roberts [1959] was to calculate a smoothed statistic from the quality deviations from target and compare it with control limits for that statistic. Under the null hypothesis of no drifting behavior and no special causes, the probability of being outside of limits is small. Therefore, when a limit violation occurs, it is assumed that a special cause is present and needs to be corrected. We continue with the notation of Box and Jenkins but restrict our attention to the one step ahead forecast ie.

$$\hat{N}_t(l) = \hat{N}_t(1) = \hat{N}_{t+1}$$

to rewrite Equation 5 as follows:

$$\hat{N}_{t+1} = \theta \, \hat{N}_t + (1 - \theta) \, N_t \tag{14}$$

where we let t equal the number of sample intervals since the last out- of-control signal. Assuming that corrective action is taken following a signal which returns the quality index to target, we set $\hat{N}_1 = 0$. Control limits are set at multiples of (usually 3x) the standard deviation of $\hat{N}_{t+1}$, which assuming an absence of drifts is

$$\sigma_{\hat{N}} = ([1 - \theta]^{2t} [(1 - \theta) / (1 + \theta)])^{\frac{1}{2}} \sigma_a$$

which converges to

$$\sigma_{\hat{N}} = [(1-\theta)/(1+\theta)]^{1/2} \sigma_a$$

after a small number of sample periods, unless q › 1. The control parameter L denotes the number of multiples of s used for the control limits, which are then set at

$$LCL = -L \, \sigma_{\hat{N}}$$

and

$$UCL = L \, \sigma_{\hat{N}}$$

Roberts and then Crowder [1987] assume that no drifts are present, and then recommend choices of q and L for controlling the average number of sample periods until an out-of-control signal for specified level shifts in the mean of the quality data away from target. For example, Crowder calculates that setting q = .95 and L = 2.5 results in an in-control average run length of 379 and an ARL of 10.8 following a level shift of $\pm 1 \, s_a$.

In this paper we recommend fitting an IMA (0,1,1) time series model to the quality data in order to determine whether or not drifts are present. This can be done with a spreadsheet program using trial and error methods to find the value of q which minimizes the forecast error variance as shown by Hunter [1986]. If no drifts are present, the estimate of q will be close to 1. Otherwise, the rate of drift is measured by the value of q as discussed previously. When this has been done, the smoothed statistic $\hat{N}_{t+1}$ gives the best forecast of the process mean and, as such, should provide a basis for making an appropriate adjustment to compensate for a drift, regardless of the setting for L. This can be illustrated by setting L equal to zero, in which case there would be an out-of-control signal for every sample. Because we have set N $= 0$ and can observe that $\hat{N}_1$ is the same as the control error, e_t , Equation 3 reduces to the following:

$$\hat{N}_2 = (1 - q)e_1$$

and the appropriate compensating adjustment with no deadtime is

$$x_1 = -(1/PG)\ (1 - q)e_1 \tag{15}$$

which is the same as Equation 11 for the time series controller which minimizes the control error variability. Therefore, avoidance of increased variability due to overcontrol is not a reason for widening the control limits (increasing L) because we have just shown that the variability is minimized with L=0. Instead, as shown by Box, Jenkins, and MacGregor [1974], the motivation for increasing L is to reduce costs when process adjustments are expensive.

The EWMA controller is then modified as follows to include the calculated adjustment:

$$\hat{N}_{t+1} = \theta\,\hat{N}_t + (1 - \theta)\,e_1$$

$$x_t = -\,(CG/PG)\,\hat{N}_{t+1} \qquad \text{if } \left|\hat{N}_{T+1}\right| > L\,\sigma_{\hat{N}}$$
$$x_t = 0 \qquad\qquad\qquad \text{otherwise.}$$

$$N_{t+1} = 0 \qquad \text{if } x_t \neq 0 \tag{16}$$

The new parameter, CG (CG > 0) is the controller gain. Setting CG = 1.0 results in an adjustment which exactly compensates for the forecasted deviation from target, and setting CG < 1 provides a convenient means of reducing the size of the adjustment.

While the performance of this EWMA controller, as measured by the control error standard deviation and the average adjustment interval, will be evaluated in this paper using simulation, it is interesting to note that an analytical approach was taken by Box and Jenkins [1963]. They used, as an example, the mass production of ball bearings, where the quality index was diameter and an adjustment to the machine required an interruption of production and hence was too costly to be made for every sample interval. They assumed that the diameters drifted away from target according to an IMA (0,1,1) process and that machine adjustments took full effect immediately. The cost of being off target was assumed to be proportional to the square of the deviation, and there was a fixed cost associated with each adjustment. In using the EWMA controller where q was chosen as above to generate the smallest forecast errors, they found that the sum of the adjustment cost and the cost associated with being off target could be minimized with the appropriate choice of L. Also, for given choices of q and L they calculated the expected run length between adjustments, E(n), and the mean squared deviation from target:

$$s_e^2 = s_a^2 + [(1-q)^2] \, g_w \, s_a^2$$

where g_w is a variance measure for a random walk process as defined by Equation 3 with $s_u = 1$, allowed to drift for E(n) periods. Table I is taken from the reference for three of their tabulated run lengths which fall within our range of interest. From it can be calculated, for any value of q from zero to one, the EWMA limit, L, and the control error standard deviation, SE = s_e/s_a. These results will be used as a check against the simulations to be discussed later.

Cumulative Sum. The Cumulative Sum (CUSUM) control procedure was originally developed by Page [1954] as a means of detecting moderate shifts in the process mean away from target more rapidly than can be done with a Shewhart chart. Early implementations used a V-Mask (Barnard [1959]) on the

Table 1. Box & Jenkins Parameters for selected Run Lengths

$E(n)$	l	g_w
7.2	2.0	.88
10.7	2.6	1.5
17.5	3.5	2.5

To Calculate EWMA Controller Performance:

$$AI = E(n)$$
$$L = l \, ((1-q)(1+q))^{1/2}$$
$$SE^2 = 1 + (1-q)^2 \, g_w$$

plotted CUSUM's in order to decide when a level change has occurred, but with the advent of computers it has become more common to use an algorithmic approach involving two one-sided CUSUM's, one for detecting level increases and the other for detecting decreases.

The CUSUM statistic is as follows:

$$S_t = \sum_{i=1}^{t} (y_i - \text{Target})$$

Figure 8 is a plot of a series of random, normally distributed data (Y) with unit standard deviation and level shifts in the mean from zero to .5 at t=25, from .5 to -1 at t=50, and from -1 to 0 at t=75. The CUSUM (S) is also plotted, and the following advantages of the CUSUM become apparent:

1. The level shifts in Y, seen as slope changes in S, are much more apparent on the CUSUM chart than on a plot of the raw data.

2. The level of Y can be estimated from the rate of change, or slope, of the CUSUM.

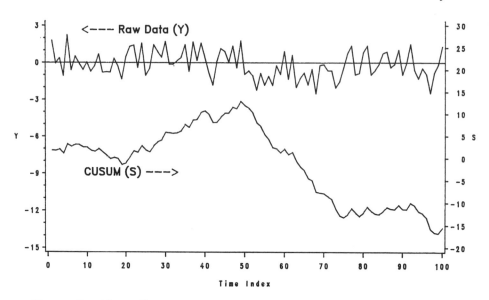

Figure 8. Normally Distributed Data with Shifts

The computational form of the CUSUM controller, as described by Lucas [1976), is as follows:

$$S_h(t) = Max[0, e_t - K + S_h(t-1)]$$
$$S_1(t) = Max [0, K - e_t + S_1(t-1)]$$

$$x_t = - (CG/PG)(K+S_h(t)/n_t) \qquad \text{if } S_h(t) >H,$$
$$x_t = -+(CG/PG)(K+S_1(t)/N_1) \qquad \text{if } S_1(t) > H,$$
$$x_t = 0 \qquad \text{otherwise.}$$

$$S_h(t) = S_1(t) = 0 \qquad \text{if } x_t \neq 0. \qquad (17)$$

In this formulation

$S_h(t)$ and $S_1(t)$ are the upper and lower CUSUMs,

$k = K/s_a$ is a tuning parameter which defines a deadband over which the controller is insensitive to level changes,

$h = H/s_a$ is another tuning parameter analogous to the spacing between the control limits on a Shewhart chart,

n_h = the number of sample periods since S (t) was last equal to zero (similar definition for n),

and PG and CG are as defined for the EWMA controller.

Lucas [1976] has evaluated the performance of the above CUSUM statistics using the average run length criterion, and it is similar to the EWMA and superior to the Shewhart chart with one set of limits at X ±3 sigma for detecting moderate shifts of the process level away from target. For example, with h = 5 and k = .5 the CUSUM ARL for a shift of one sigma is 10.4 while the in-control ARL is 465. The corresponding ARL's for the conventional Shewhart chart are 44 and 370.4, respectively. Additionally, this CUSUM controller provides a basis for adjustment because the slope of the CUSUM and hence, the new process level is estimated by $K + S_h(t)/n$ or $-(K + S_1(t)/n_1)$. There is no prior work, however, which estimates the control error sigma or the average adjustment interval when the disturbances follow an IMA (0,1,1) process.

Methodology For Simulation Study

Controller Performance Measures. The performance measures for these evaluations are (1) SE, the control error standard deviation expressed as a multiple of s , the standard deviation of the random noise component of the process disturbance, and (2) AI, the average number of sample periods between adjustments. An optimal controller is one which, for any average adjustment interval (AI), value of r, and transfer function, gives the lowest process variability (SE). Once these controllers have been found, which we will accomplish via simulation, the adjustment interval can be varied in order to minimize the sum of adjustment and out-of- specification costs.

Simulation Program Design. The objective of this work was accomplished by writing simulation programs for the IMA disturbance, the process dynamic response, and each candidate feedback control algorithm. These programs use SAS Software, and separate program listings for the EWMA and CUSUM controllers are included as Appendices 1 and 2. The random shocks, a , were obtained from the SAS function RANNOR with the seed based on the clock time in the computer. This helped assure complete randomization of the simulation runs and prevented any bias in the study results. The programs were run on a Digital Equipment Corporation Vax-Station 2000 running the VMS operating system. Simulation of Time Series control was not necessary because, as we have seen in the "Time Series Controllers" section, the performance of this controller for the cases under study is known.

Experimental Strategy for Simulation Studies. There are three parameters which determine the process simulation: b, the number of full periods of deadtime; w, a measure of the amount of process response carrying over into additional sample periods; and r, a measure of the rate of drift caused by process or external disturbances. Without loss of generality, we can set the first period process response to a unit step change (1) and the standard deviation of the random shocks (s) equal to unity. The emphasis for this work is on control of quality indices generated in the laboratory from periodic production samples. Cost factors usually result in a sample interval which is large in relation to the dynamic response of the process to an adjustment. However, there are often cases where delays in performing and reporting the analysis result in a full period of deadtime. For these reasons, we limit our study to the following six cases:

Case #	d	b	q
1	0	0	.25
2	0	0	.50
3	0	0	.75
4	0	1	2 5
5	0	1	.50
6	0	1	.75

The controller tuning parameters are L and CG for the EWMA controller, assuming we set q to match the IMA disturbance; and h, k, and CG for the CUSUM controller. Rather than presenting simulation results for arbitrary sets of controller tuning parameters, we restrict our performance comparisons to tuning parameter combinations which are optimal in the sense of giving nearly minimal control error sigmas for a given

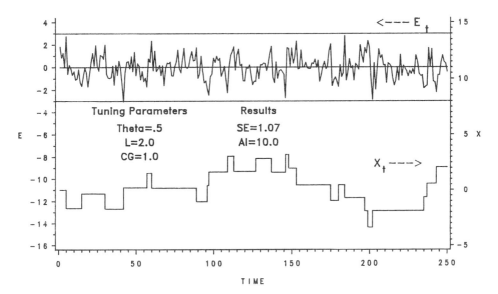

Figure 9. Simulation of EWMA Control with Optimal Tuning

average adjustment interval. This at once eliminates over-control which is characterized by more variable control errors and more frequent adjustments. This is shown by Figures 9 and 10, which compare the performance of an optimally tuned EWMA controller with one having CG set too high.

Response surface experiments were run for both controllers on the six above cases in order to determine optimal tuning parameter combinations. The two responses of interest were the control error sigma (SE) and the adjustment frequency (AF). The tuning parameters served as experimental factors. Each experiment consisted of a simulation run of 2000 sample intervals. The experimental design used was Central Composite Design with the relative spacing of the star and factorial points and the number of center points set to give uniform precision as discussed by Myers [1971]. Empirical models of the following form were fit to the data in each of the 12 experiments using least squares regression analysis:

$$Y = b_0 + b_1 X_1 + \ldots\ldots b_k X_k + b_{11} X_1^2 + \ldots. b_{kk} X_k^2 + b_{12} X_1 X_2 + \ldots\ldots + b_{(k-1)k} X_{k-1} X_k$$

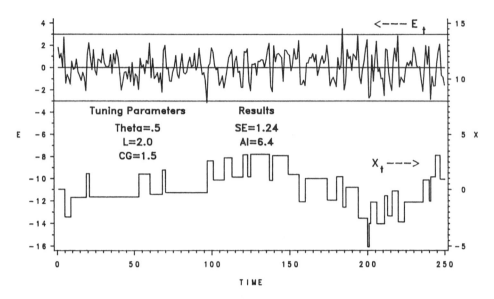

Figure 10. Simulation of EWMA Control with Too Much Gain

where X_1 to X_k are the experimental factors and b_i, b_{ii}, and b_i
$(i+1)$ are least squares parameter estimates. Then an
optimization procedure based on the Nelder-Mead Simplex
Search Algorithm (Nash [1979]) was used to find tuning
parameter combinations giving minimal control error sigma
subject to an upper constraint on adjustment frequency. By
varying these constraints a series of optimal parameter
combinations was generated covering a range of adjustment
intervals from 3 to 20. Then, an additional simulation run of
10,000 sample periods was made for each optimal set of tuning
parameters in order to more precisely estimate the controller
performance.

Single Case Example of Finding Optimal EWMA Controllers.
Instead of presenting the experimental results for both
controllers and all six cases, we use Case #2 for the EWMA
controller of Equation16 to illustrate the methodology. The
experimental design, which is shown in Table 2 and Table 3,
gives the the run order and the simulation results for all items.
The output of the simulation program for Item #1 is shown on
Table 4, which gives the standard deviation of the control errors

Table 2.   Experimental Design for EWMA Case #2 Optimization

# Reps	L	CG
1	1.25	0.65
1	3.08	0.65
1	1.25	1.35
1	3.08	1.35
1	0.87	1.00
1	3.46	1.00
1	2.16	0.50
1	2.16	1.50
5	2.16	1.00

Table 3.   Worksheet for Case #2 Optimization

Run #	L	CG	SE	AF
1	2.16	1.00	1.171	0.110
2	3.46	1.00	1.328	0.043
3	2.16	1.00	1.126	0.093
4	2.16	1.50	1.200	0.115
5	3.08	0.65	1.333	0.066
6	1.25	0.65	1.119	0.257
7	0.87	1.00	1.035	0.353
8	2.16	0.50	1.178	0.113
9	2.16	1.00	1.172	0.105
10	2.16	1.00	1.150	0.105
11	3.08	1.35	1.330	0.065
12	2.16	1.00	1.145	0.102
13	1.25	1.35	1.076	0.241

Table 4.   EWMA Controller Simulation

Theta=.50	L=2.16	CG=1.0	Delta=.000	b=0

VARIABLE	N	MEAN	STANDARD DEVIATION
ET	2000	0.021	1.171
			=====
DXT	2000	-0.002	0.513
FREQ	2000	0.110	0.313
		=====	

(e_t) and an average for an indicator variable (FREQ), which takes the value 1 for sample periods with an adjustment and zero otherwise. The reciprocal of this average is the average adjustment interval. Figure 11 is a time series plot of the controlled and manipulative variable for the first 250 runs.

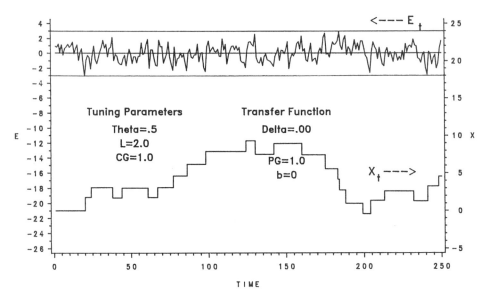

Figure 11. Simulation of EWMA Control for Case #2 - First 250 Runs

Table 5.   Regression Coefficients for SE

Source Term	Expanded Term	Coeff. Value	Coeff. Stderr.	Signif.
1	1	1.153	0.011	0.0001
L	((L-(2.163))/1.297)	0.156	0.012	0.0001
CG	((G-(1))/0.5)	-0.003	0.012	0.8384
L**2	((L-(2.163))/1.297)**2	0.043	0.018	0.0485
CG**2	((G-(1))/0.5)**2	0.051	0.019	0.0259

R-SQUARE = 0.9566   Adjusted R-SQUARE = 0.9348
STD. ERR. in SE = 0.024323   Deg. Freedom = 8

A stepwise regression analysis was performed on the data. Terms were removed from the model if their contribution to the fit was not significant at a= 10, and could not be re-entered unless significant at a = .05. Tables 5 and 6 show the coefficients of the fitted models for SE and AF, and Tables 7 and 8 are the Analysis of Variance tables. It can be seen that the models for SE and AF explain 96 and 99% of the variability among the simulation runs, and that there is no strong evidence of lack of fit. Contour plots of AI=1/AF and SE versus L and CG are shown on Figure 12, and they indicate that for any of the AI

Table 6.   Regression Coefficients for AF

Source Term	Expanded Term	Coeff. Value	Coeff. Stderr.	Signif.
1	1	0.107	0.004	0.0001
L	((L-(2.163))/1.297)	-0.143	0.005	0.0001
L**2	((L-(2.163))/1.297)**2	0.095	0.008	0.0001

R-SQUARE = 0.9883   Adjusted R-SQUARE = 0.9859
STD. ERR. in AF = 0.01072   Deg. Freedom = 10

Table 7.   ANOVA TABle for SE

Source	df	Sum Sq.	Mean Sq.	F-Ratio	Sig.Lev.
Total(Corr)	12	0.109	0.009		
Regression	4	0.104	0.026	44.030	0.000
Residual	8	0.005	0.001		
Lack of Fit	4	0.003	0.001	2.183	0.234
Pure Error	4	0.001	0.000		

R-SQUARE = 0.9566   Adjusted R-SQUARE = 0.9348

F(4,8) as large as 44.03 is a very rare event =>
   highly unlikely that all coefficients are zero.
F(4,4) as large as 2.183 is not a rare event =>
   no evidence of lack of fit.
Estimate of Pure Error from 1 group of replicates.

Table 8.   ANOVA Table for AF

Source	df	Sum Sq.	Mean Sq.	F-Ratio	Sig.Lev.
Total(Corr)	12	0.098	0.008		
Regression	2	0.097	0.048	421.000	0.000
Residual	10	0.001	0.000		
Lack of Fit	6	0.001	0.000	4.182	0.094
Pure Error	4	0.000	0.000		

R-SQUARE = 0.9883   Adjusted R-SQUARE = 0.9859

F(2,10) as large as 421 is a very rare event =>
   highly unlikely that all coefficients are zero.
F(6,4) as large as 4.182 is a moderately rare event =>
   some evidence of lack of fit.
Estimate of Pure Error from 1 group of replicates.

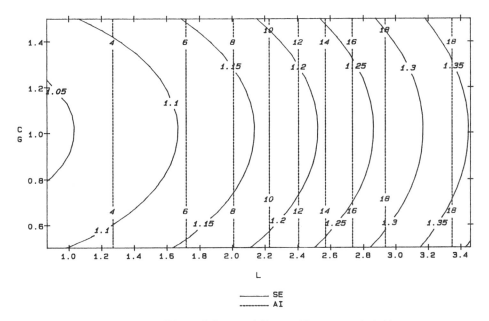

Figure 12. Contour Plot of Control Error Sigma and Adjustment
Interval for Case #2

Table 9.   Case 2, EWMA Simulation Results

Theta	b	L	CG (NM*)	SE (NM)	AF (NM)	CG (Ext**)	SE (Ext)	AF (Ext)	AI (Ext)
0.5	0	0.50	1.02	1.041	0.353	1.00	1.030	0.363	2.755
0.5	0	0.83	1.02	1.075	0.202	1.00	1.076	0.190	5.263
0.5	0	1.00	1.08	1.101	0.149	1.00	1.089	0.140	7.143
0.5	0	1.15	1.02	1.130	0.122	1.00	1.137	0.117	8.547
0.5	0	1.31	1.03	1.163	0.094	1.00	1.186	0.102	9.804
0.5	0	1.48	1.03	1.197	0.077	1.00	1.210	0.079	12.658
0.5	0	1.67	1.02	1.251	0.065	1.00	1.269	0.066	15.152
0.5	0	1.83	1.03	1.297	0.055	1.00	1.311	0.056	17.857

*   NM = Results from Nelder-Mead Optimizations
** Ext = Results from extended runs of 10,000 samples

contours the control error sigma is lowest when the controller gain is about 1. Table 9 shows sample results of the Nelder-Mead optimizations along with the results of the extended simulations. The optimum controller gain is found to be near 1, and because making an adjustment which exactly compensates for the forecasted deviation has intuitive appeal for the zero deadtime case, CG was set to 1.0 for the extended runs in this case. Figure 13 is a scatterplot of SE versus AI for the extended runs at optimal settings (CG = 1.0) and the experimental data for other values of CG. It shows that the optimal controllers lie along the lower edge of the scatter. Complete listings of the results of extended simulations are given in Tables 10 to 13 for the optimal EWMA and CUSUM controllers with zero and one period of deadtime.

Generalized EWMA Performance Model. Figure 14 is a scatterplot of SE vs AI for data from all six EWMA cases. Superimposed on these data are curves from the following regression model:

$$SE = 1 + g [b_0 I + b_1 (AI-1) + b_2 (AI - 1)^2] \qquad (18)$$

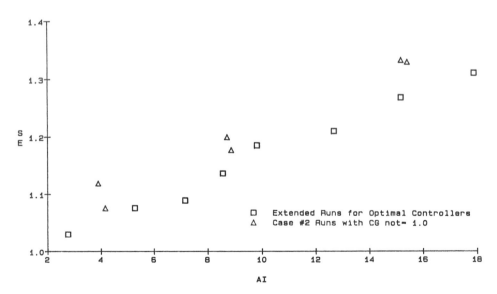

Figure 13. Performance Comparison of Optimal vs Non-Optimal EWMA Controllers for Case #2

Table 10. Extended Simulation Runs for EWMA Controller

Theta	b	L	CG	SE	AF	AI
0.25	0	0.97	1.00	1.04	0.352	2.8
0.25	0	1.19	1.00	1.09	0.284	3.5
0.25	0	1.38	1.00	1.12	0.240	4.2
0.25	0	1.64	1.00	1.17	0.188	5.3
0.25	0	1.96	1.00	1.22	0.141	7.1
0.25	0	2.18	1.00	1.27	0.121	8.3
0.25	0	2.49	1.00	1.34	0.100	10.0
0.25	0	2.85	1.00	1.43	0.079	12.7
0.25	0	3.23	1.00	1.54	0.065	15.4
0.25	0	3.87	1.00	1.64	0.046	21.7
0.50	0	0.87	1.00	1.03	0.363	2.8
0.50	0	1.44	1.00	1.08	0.190	5.3
0.50	0	1.73	1.00	1.09	0.140	7.1
0.50	0	1.99	1.00	1.14	0.117	8.5
0.50	0	2.27	1.00	1.19	0.102	9.8
0.50	0	2.56	1.00	1.21	0.079	12.7
0.50	0	2.89	1.00	1.27	0.066	15.2
0.50	0	3.17	1.00	1.31	0.056	17.9
0.75	0	0.79	1.00	1.01	0.292	3.4
0.75	0	1.06	1.00	1.02	0.198	5.1
0.75	0	1.32	1.00	1.05	0.155	6.5
0.75	0	1.59	1.00	1.04	0.110	9.1
0.75	0	1.85	1.00	1.05	0.088	11.4
0.75	0	2.12	1.00	1.07	0.069	14.5
0.75	0	2.38	1.00	1.08	0.056	17.9
0.75	0	2.65	1.00	1.08	0.045	22.2
0.75	0	3.31	1.00	1.14	0.030	33.3

Table 11. Extended Simulation Runs for EWMA Controller

Theta	b	L	CG	SE	AF	AI
0.25	1	1.06	0.53	1.32	0.423	2.4
0.25	1	1.42	0.50	1.39	0.320	3.1
0.25	1	1.63	0.53	1.42	0.276	3.6
0.25	1	1.85	0.57	1.42	0.220	4.5
0.25	1	2.10	0.63	1.48	0.177	5.6
0.25	1	2.40	0.67	1.56	0.156	6.4
0.25	1	2.76	0.73	1.59	0.111	9.0
0.25	1	3.11	0.79	1.66	0.091	11.0
0.25	1	3.63	0.83	1.77	0.065	15.4
0.50	1	1.02	0.84	1.16	0.345	2.9
0.50	1	1.39	0.85	1.19	0.238	4.2
0.50	1	1.61	0.85	1.20	0.188	5.3
0.50	1	1.78	0.85	1.20	0.154	6.5
0.50	1	1.94	0.85	1.23	0.142	7.0
0.50	1	2.06	0.85	1.25	0.131	7.6
0.50	1	2.27	0.85	1.29	0.111	9.0
0.50	1	2.49	0.85	1.27	0.088	11.4
0.50	1	2.84	0.85	1.38	0.078	12.8
0.50	1	3.12	0.85	1.37	0.059	16.9
0.75	1	1.03	1.00	1.05	0.220	4.5
0.75	1	1.14	1.00	1.06	0.195	5.1
0.75	1	1.40	1.00	1.07	0.140	7.1
0.75	1	1.53	1.00	1.08	0.127	7.9
0.75	1	1.75	1.00	1.09	0.100	10.0
0.75	1	2.33	1.00	1.10	0.058	17.2
0.75	1	2.59	1.00	1.12	0.051	19.6
0.75	1	3.15	1.00	1.15	0.035	28.6

Table 12. Extended Simulation Runs for CUSUM Controller

Theta	b	h	k	CG	SE	AF	AI
0.25	0	0.80	0.55	1.10	1.12	0.281	3.6
0.25	0	1.00	0.45	1.10	1.12	0.274	3.6
0.25	0	1.00	0.69	1.10	1.19	0.217	4.6
0.25	0	1.29	0.78	1.00	1.22	0.163	6.1
0.25	0	1.13	1.00	1.10	1.26	0.147	6.8
0.25	0	1.51	1.00	1.00	1.31	0.127	7.9
0.25	0	1.51	1.00	1.10	1.30	0.121	8.3
0.25	0	2.00	1.00	1.00	1.34	0.105	9.5
0.25	0	2.00	1.20	1.00	1.42	0.085	11.8
0.50	0	1.05	0.39	0.72	1.06	0.258	3.9
0.50	0	1.20	0.51	0.80	1.10	0.202	5.0
0.50	0	1.68	0.50	1.01	1.14	0.158	6.3
0.50	0	1.79	0.67	1.00	1.17	0.121	8.3
0.50	0	2.35	0.72	1.04	1.21	0.095	10.5
0.50	0	2.26	0.86	0.95	1.24	0.083	12.0
0.50	0	2.76	0.98	0.83	1.30	0.067	14.9
0.50	0	3.20	1.00	0.85	1.32	0.059	17.1
0.50	0	3.33	1.00	1.00	1.33	0.053	18.9
0.75	0	1.80	0.20	0.55	1.03	0.183	5.5
0.75	0	3.00	0.00	0.75	1.05	0.153	6.5
0.75	0	3.55	0.00	0.71	1.05	0.131	7.6
0.75	0	4.31	0.00	0.91	1.05	0.104	9.6
0.75	0	3.00	0.30	0.51	1.07	0.095	10.5
0.75	0	4.71	0.04	0.85	1.07	0.092	10.9
0.75	0	4.28	0.19	0.76	1.08	0.080	12.5
0.75	0	5.00	0.50	0.75	1.15	0.046	21.7
0.75	0	4.20	0.80	0.96	1.16	0.033	30.3

where

$$g = [1 + (1-q)^2]^{1/2} - 1 \qquad (19)$$

b_0 , b_1 , and b_2 are least squares parameter estimates with values,

$$b_0 = 1.14$$
$$b_1 = .166$$
$$b_2 = -.0018;$$

Table 13. Extended Simulation Runs for CUSUM Controller

Theta	b	h	k	CG	SE	AF	AI
0.25	1	2.04	0.50	0.75	1.50	0.218	4.6
0.25	1	2.40	0.50	0.75	1.51	0.193	5.2
0.25	1	2.69	0.55	0.66	1.53	0.172	5.8
0.25	1	3.17	0.50	0.75	1.56	0.156	6.4
0.25	1	2.94	0.72	0.71	1.58	0.137	7.3
0.25	1	3.30	0.84	0.78	1.66	0.120	8.3
0.25	1	3.64	0.97	0.77	1.73	0.105	9.5
0.50	1	2.92	0.03	0.62	1.25	0.208	4.8
0.50	1	3.34	0.03	0.65	1.27	0.187	5.3
0.50	1	3.99	0.07	0.71	1.31	0.158	6.3
0.50	1	3.45	0.36	0.73	1.30	0.121	8.3
0.50	1	3.09	0.58	0.71	1.33	0.107	9.3
0.50	1	2.43	0.85	0.69	1.35	0.098	10.2
0.50	1	3.51	0.91	0.83	1.40	0.067	14.9
0.50	1	4.00	1.00	0.85	1.45	0.056	18.0
0.50	1	4.50	1.00	0.85	1.48	0.053	18.7
0.75	1	3.20	0.04	0.62	1.11	0.142	7.0
0.75	1	3.49	0.08	0.59	1.08	0.116	8.6
0.75	1	3.92	0.11	0.79	1.11	0.104	9.6
0.75	1	4.13	0.20	0.94	1.10	0.081	12.3
0.75	1	3.98	0.36	1.04	1.13	0.064	15.6
0.75	1	3.60	0.73	0.46	1.18	0.044	22.7
0.75	1	3.87	0.81	0.98	1.21	0.037	27.0
0.75	1	3.47	0.84	0.61	1.18	0.037	27.0

and I is an indicator variable taking the value zero if there is no deadtime or one if a period of deadtime exists in the feedback loop. The RSQUARED statistic for this model is .99, meaning that the model explains 99% of the variation in the data. Therefore it is useful for evaluating the effects of deadtime and adjustment interval on the control error sigma for any value of q.

The form of the model was determined by observing that for any adjustment interval, the slope of the relationship between SE and AI varies with q in the same manner as the fractional increase in SE caused by one period of deadtime. Both increase as the rate of drift increases, i.e. as q gets closer to zero. q is

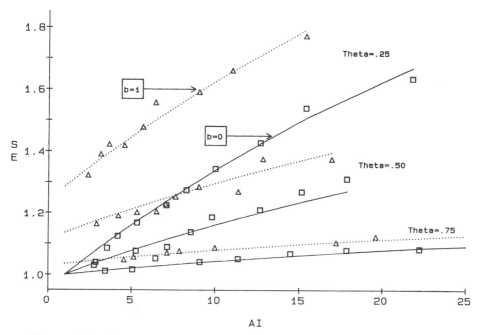

Figure 14. Scatter Plot of EWMA Simulation Data Compared with
Performance Model Curves

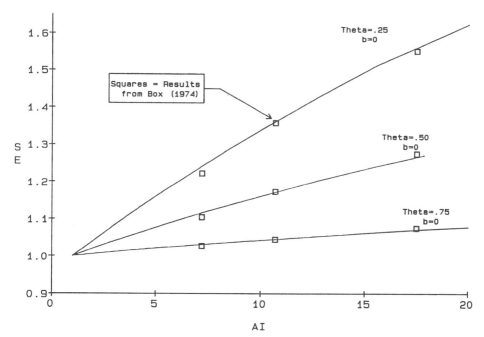

Figure 15. Comparison of EWMA Performance Model Predictions
with Results from BOX (1974)

taken from Equation 8 with P=2, and it can be described as the fractional increase in the error sigma for an optimal time series controller caused by adding the first period of deadtime. As a further check on the model, Figure 15 compares its predictions with those based on Table I from the work of Box, Jenkins, and Macgregor [1974]. It is observed that there is good agreement with this model.

Discussion of Simulation Results

<u>Fitted Model for EWMA Control.</u> The first aspect of the fitted EWMA model is that, for each of the six cases studied, the control error sigma increases as the average adjustment interval increases. Those controllers with a higher average adjustment interval have a larger L, hence the process is allowed to drift farther from target before an adjustment is made, causing an increase in the control error sigma. Figures 16 and 17 compare the performance of EWMA controllers with L=3 and L=1.5 for Case 2 (q = .5, no dead time). The one-step-ahead

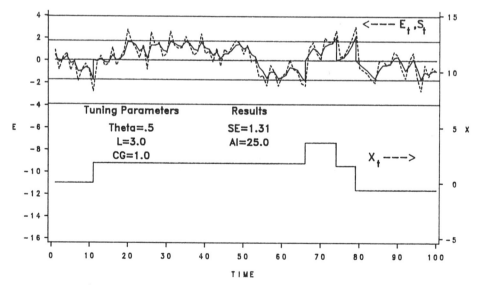

Figure 16. Simulation of EWMA Control with 3-Sigma Limits

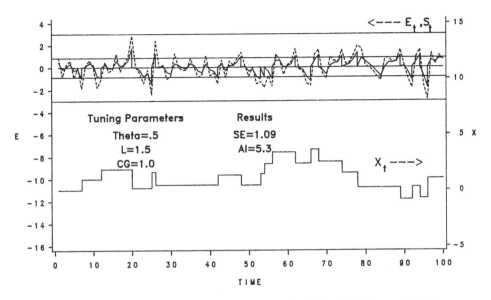

Figure 17. Simulation of EWMA Control with 1.5-Sigma Limits

forecasts are superimposed on the raw process variable data along with the EWMA limits at

$$\pm L \ [(1-q)/(1+q)]^{1/2}$$

The lower control error sigma for L=1.5 is seen to be the result of more frequent intervention with smaller adjustments to return the process to target. The performance for L=3 is typical of using a Statistical Process Control strategy with an ARL of 400 assuming the process is on target with no drifts. Fitted normal density curves of e for these two controllers (Figure 18) show the greater concentration near target associated with the tighter EWMA limits.

It is interesting that for no dead time (b=0) L is the only tuning parameter for the EWMA controller, because optimal control in the above sense is obtained with CG = 1.0 and q is determined from analysis of the open loop data as discussed in Section 3.2. The choice of L is purely a matter of economics, rather than being based on statistical considerations. For example, if the adjustments are automated and cost nothing,

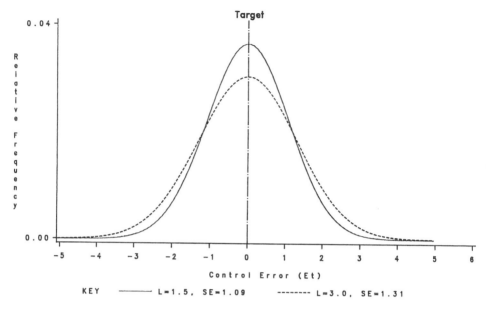

Figure 18. Effect of Tighter EWMA Limits on Control Error Sigma

and if specification limits are narrow relative to process variability, then it is best to set L=0 resulting in the time series controller as shown by Equations 11 and 15 in order to minimize the production of off-quality material. On the other hand, if adjustments are not automated, if one operator is responsible for a large array of machines, and if specification limits are wide, then it is best to set L=3 in order to minimize adjustment costs.

The next aspect of the EWMA performance model of Equation 18 and Figure 14 deserving mention, is the effect of changes in q on the relationship between control error sigma and adjustment interval. This is shown by the contour plots of SE versus q and AI of Figures 19 and 20 based on this model for the cases of zero and one period of dead- time, respectively. A flatter surface is seen toward the right hand side for large r. If one follows a contour of constant control error sigma, it is seen that the adjustment interval has a non-linear relationship with q with more frequent adjustments for lower values of r. For example, with no deadtime a sigma of 1.1 can be achieved with one adjustment for every 30 samples if q is .75, but the

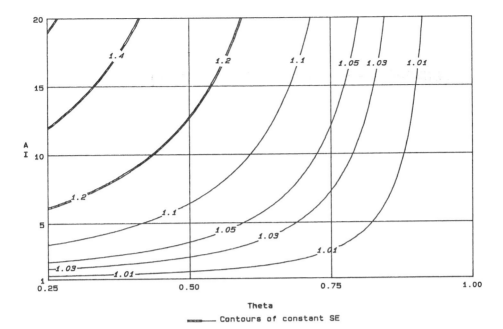

Figure 19. Contour Plot of Control Error Sigma vs Theta and
Adjustment Interval for No Deadtime

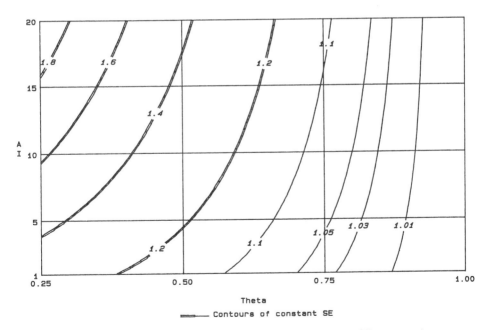

Figure 20. Contour Plot of Control Error Sigma vs Theta and
Adjustment Interval for One Period of Deadtime

intervals shorten to 7 and 3 for q of .5 and .25 respectively. Therefore, if the process has a small q (fast drifts), then steps should be taken to reduce the adjustment cost.

The final aspect of interest from the EWMA performance model in Equation 18, is the effect of deadtime on the control error sigma. For a time series controller, the presence of deadtime means that the process disturbances must be forecasted farther ahead in order to determine the best adjustment, and, as shown in Equation 8, this results in a larger forecast error sigma and hence a larger control error sigma. The table below compares the control error sigmas (expressed as a multiple of s_a) for the time series controller of Equation 12 and the optimal EWMA controller with L=0.

Controller Performance with One Deadtime Period

q	SE (Time Series)	SE (EWMA, L=0)
.75	1.031	1.035
.50	1.118	1.135
.25	1.250	1.285

Remembering that SE for all of the above cases with no deadtime is 1.0, the penalty for having one period of deadtime in the feedback loop is seen to be more severe for processes with fast drifts. Also, the EWMA controller is somewhat more severely affected than the time series controller resulting from the fitted value of b_0 =1.145 in Equation 18 rather than 1.0. This is because the EWMA controller has no deadtime compensation term and as shown in Table XI, requires a controller gain below one in order to avoid over-control.

Additionally, it is noted that the zero and one deadtime curves for each value of q on Figure 14 are parallel, resulting from lack of an interaction or cross-product term of the form I(AI-1) or I(AI-1)^2. Therefore, the amount of increase in sigma resulting from one period of deadtime is the same for an EWMA controller with L=3 as it is for L=0. In other words, the amount of increase in sigma caused by extending the adjustment interval is the same whether or not there is a period of deadtime. Therefore, there is a benefit resulting from the elimination of deadtime even if one is entertaining an EWMA controller with infrequent adjustments.

<u>CUSUM versus EWMA Control</u>. Figures 21 and 22 compare the extended CUSUM simulation results with the EWMA model predictions for the zero and one deadtime period cases, respectively. It can be seen that the CUSUM results have a higher control error sigma than the corresponding EWMA results for all cases studied. Figure 23 contains histograms of actual CUSUM and EWMA results minus the EWMA predictions, showing that the average sigma increase from using CUSUM instead of EWMA is .04.

The likely reason for this small, but statistically significant difference in performance, is that the adjustments based on the EWMA forecast do a better job of returning the process to target

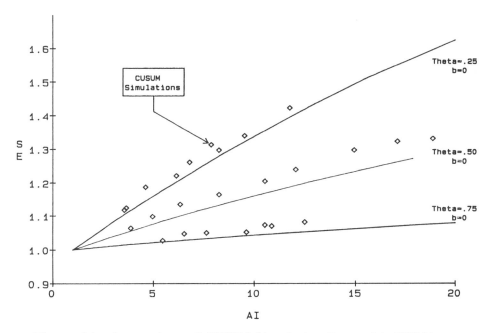

Figure 21.   Comparison of CUSUM Simulation Data with EWMA Performance Model Curves with No Deadtime

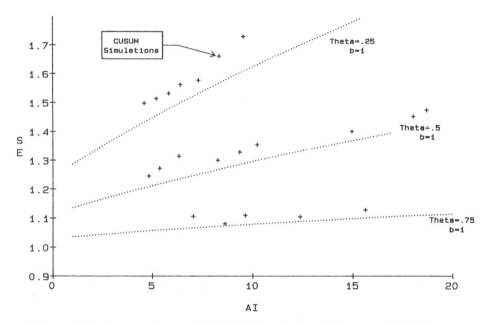

Figure 22. Comparison of CUSUM Simulation Data with EWMA
           Performance Model Curves with One Period of Deadtime

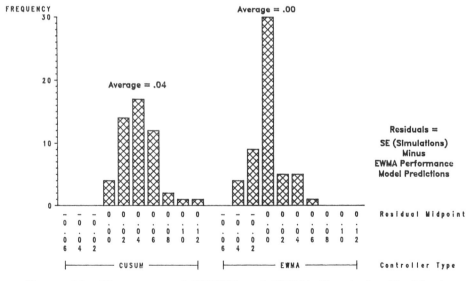

Figure 23. Histograms of CUSUM and EWMA Simulation Residuals vs.
           the EWMA Performance Model

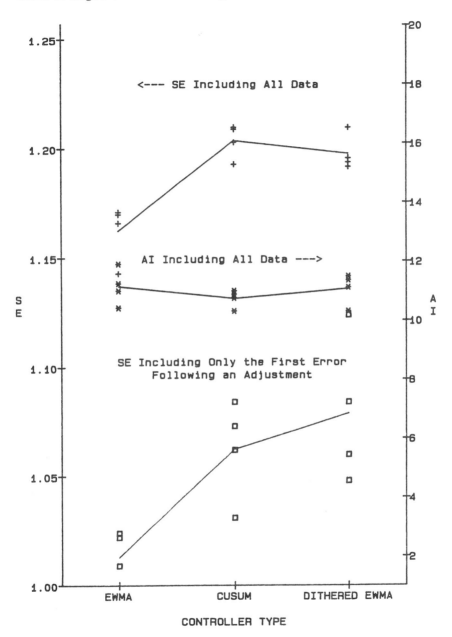

Figure 24. Effect of Adjustment Errors on Control Error Sigma (SE)
with Constant Adjustment Interval

than the CUSUM adjustments based on Equation 17 when the process disturbance can be represented by an IMA(0,1,1) model. As evidence of this, four simulation runs of 5000 sample intervals were made using Case #2 (q = 5, no deadtime) for each of three controller items:

Item 1:   EWMA with L=2.3, adjustment errors (control errors at sample intervals immediately following adjustments) with a sigma of 1.01 in close agreement with the EWMA one step ahead forecast error sigma of 1.0.

Item 2:   EWMA with L=2.38 and a dither action to cause adjustment errors with approximately the same sigma as that for the CUSUM controller.

Item 3:   CUSUM with h=2.35 and k=.72 (determined to be optimal from prior experimentation but found to have a sigma of 1.07 for adjustment errors).

The results plotted in Figure 24 show that when the EWMA controller is modified in order to have the same adjustment interval and adjustment error sigma as the CUSUM controller, it has approximately the same control error sigma for all samples. Therefore, it is concluded that precision of the estimator for new process level accounts for most of the difference in performance between the CUSUM and EWMA controllers.

The fact that the CUSUM estimator for the new process level is more variable than the EWMA forecast is not surprising, because the EWMA forecasts are the minimum variance forecasts for the cases studied, while the CUSUM estimator is based on the premise of a level shift occurring n_h (or n_l) periods ago. It might be argued that the relative performance of these two controllers should be evaluated with a process disturbance consisting of white noise superimposed onto randomly occurring step changes. This has not been done, however, because the objective of this paper is to study control algorithms for drifting processes.

References

Bagshaw, M. and Johnson, R. J. (1974) - "The Effect of Serial Correlation on the Performance of CUSUM Tests", Technometrics 16, pp. 103-112.

Bagshaw, M. and Johnson, R. J. (1975) - "The Effect of Serial Correlation on the Performance of CUSUM Tests II", Technometrics 17, pp. 73-80.

Barnard, G. A. (1959) - "Control Charts and Stochastic Processes", Journal of the Royal Statistical Society, Series B 21, pp. 239-271.

Bissell, A. F. (1984) - "The Performance of Control Charts and Cusums under Linear Trend", Applied Statistics 33, No. 2, pp. 145- 151.

Box, G. E. P. and Jenkins, G. M. (1963) - "Further Contributions to Adaptive Quality Control: Simultaneous Estimation of Dynamics: Non- Zero Costs", Bull. Int. Statist. Inst., 34th Session, pp. 943-974.

Box, G. E. P. and Jenkins, G. M. (1976) Time Series Analysis: Forecasting and Control. Holden Day, Inc., San Francisco, CA.

Box, G. E. P., Jenkins, G. M., and MacGregor, J. F. (1974) - "Some Recent Advances in Forecasting and Control, Part II", Applied Statistics 23, No. 2, pp. 158-179.

Champ, C. W. and Woodall, W. H. (1987) - "Exact Results for Shewhart Control Charts with Supplementary Runs Rules", Technometrics 29, No. 4, pp. 393-399.

Crowder, S. V. (1987) - "A Simple Method for Studying Run-Length Distributions of Exponentially Weighted Moving Average Charts", Technometrics, 29, No. 4, pp.401-407.

Ford Motor Company (1984) - Continuing Process Control and Process Capability Improvement.

Hunter, J. S. (1986) - "The Exponentially Weighted Moving Average", Journal of Quality Technology 18, No. 4, pp. 203-210.

Kane V. E. (1986) - "Process Capability Indices", Journal of Quality Technology 18, No. 1, pp. 41-52.

Lucas, J. M. (1976) - "The Design and Use of V-Mask Control Schemes", Journal of Quality Technology 8, No. 1, pp. 1-12.

Lucas, J. M. and Sacucci, M. S. (1987) - "Exponentially Weighted Moving Average Control Schemes: Properties and Enhancements", Faculty Working Paper, Drexel University, College of Business and Administration, Philadelphia, PA.

Myers, R. H. (1971) - Response Surface Methodology, Allyn and Bacon, Inc., Boston, Mass., p. 153.

Nash, J. C. (1979) - Compact Numerical Methods for Computers: Linear Algebra and Function Minimization, Adam Hilger Ltd., Bristol, Eng.

Page, E. S. (1954) - "Continuous Inspection Schemes", Biometrica 41, pp. 100-115.

Palmor, Z. J. and Shinnar, R. (1979) - "Design of Sampled Data Controllers", Ind. & Eng. Chem., Process Des. & Dev. 18, No. 1, pp. 8-30.

Roberts, S. W. (1959) - "Control Chart Tests Based on Geometric Moving Averages", Technometrics 1, No. 3, pp. 239-250.

Shewhart, W. A. (1931) - Economic Control of Quality of Manufactured Product, D. Van Nostrand Co., Inc. New York, N. Y.

Smith, O. J. M. (1959) -   ISA Journal 6, No. 2, p. 28.

Western Electric Co., Inc. (1956) - Statistical Quality Control Handbook, Delmar Printing Co., Charlotte, N. C.

Appendix 1

Box & Jenkins Parameters for Selected Run Lengths

$E(n)$	l	g_w
7.2	2.0	8 8
10.7	2.6	1.5
17.5	3.5	2.5

To Calculate EWMA Controller Performance:

$$AI = E(n)$$

$$L = 1 \ ((1-q)(1+q))^{1/2}$$

$$SE^2 = 1 + (1-q)^2 \ g_w$$

EWMA SIMULATION PROGRAM

THIS PROGRAM SIMULATES AN EWMA CONTROLLER WITH A FIRST ORDER PLUS DEAD- TIME TRANSFER FUNCTION AND AN IMA(0,1,1) DISTURBANCE PROCESS. IT RUNS ON VMS WITH SAS/BASIC AND SAS/GRAPH SOFTWARE, VERSION 5.16. A1=delta, and B0-B3 are parameters used to change omega and b. DB is the same as the EWMA parameter, L. THE GMA THETA IS CONSTRAINED TO BE THE SAME AS THETA FOR THE DISTURBANCE PROCESS.

Different case studies are performed by changing the numbers in the following %LET statements. It is convenient to run the program from the full screen display (FSD) after removing this comment.;

```
OPTIONS DQUOTE ls=78;
%LET SPROC=1  ;  %LET THETA=.75; %LET DB=2.160; %LET CG=1.0 ; %LET
A1=.000; %LET B0=1  ; %LET B1=0  ; %LET B2=0  ; %LET B3=0 ;
DATA A;

DO TIME=1 TO 2000;T=0;
RETAIN ATM1 NTM1 GMATM1 XTM4 XTM3 XTM2 XTM1 YTM1 0;
*************  DISTURBANCE   PROCESS *****************;
AT=&SPROC*RANNOR(0);
NT=NTM1+AT-&THETA*ATM1;
***************** TRANSFER  FUNCTION  *******************;
YT=&A1*YTM1+&B0*XTM1+&B1*XTM2+&B2*XTM3+&B3*XTM4;
ET=YT+NT;
PG=(&B0+&B1+&B2+&B3)/(1-&A1);
******************** CONTROLLER *******************;
GMAT=(1-&THETA)*ET+&THETA*GMATM1;
IF GMAT LT -&DB*SQRT((1-&THETA)/(1+&THETA))
OR GMAT GT &DB*SQRT((1-&THETA)/(1+&THETA))
THEN DXT=-&CG*GMAT/PG;ELSE DXT=0;
```

```
IF GMAT LT -&DB*SQRT((1-&THETA)/(1+&THETA))
OR GMAT GT   &DB*SQRT((1-&THETA)/(1+&THETA)) THEN GMAT=0;
XT=XTM1+DXT;
***************** UPDATE VARIABLES ***************;
IF DXT NE 0 THEN FREQ=1;ELSE FREQ=0;
ATM1=AT;
GMATM1=GMAT;
NTM1=NT;
XTM4=XTM3;
XTM3=XTM2;
XTM2=XTM1;
XTM1=XT;
YTM1=YT;


********** DROP UNNEEDED VARIABLES FOR PRINTOUT ******;
DROP ATM1 NTM1;
DROP XTM1 YTM1;
OUTPUT;END;
****************** PROCEDURES ********************;
PROC MEANS MAXDEC=3;VAR ET DXT FREQ;
TITLE 'EWMA CONTROLLER SIMULATION';
TITLE3 "SPROC=&SPROC  THETA=&THETA  DB=&DB  CG=&CG"; TITLE4
"DELTA=&A1 B0=&B0  B1=&B1  B2=&B2  B3=&B3"; DATA B;SET A;IF TIME
LE 250;
PROC GPLOT;
PLOT (ET T)*TIME/OVERLAY
VAXIS=-26 TO 4 BY 2 HAXIS=0 TO 250 BY 50
VREF=-3,3 CV=RED CAXIS=YELLOW CTEXT=CYAN;
PLOT2 XT*TIME/
     VAXIS=-10 TO 30 BY 5 CAXIS=YELLOW CTEXT=ORANGE; SYMBOL1
I=JOIN C=CYAN ;
SYMBOL2 I=JOIN C=GREEN;
SYMBOL3 I=JOIN C=ORANGE;
TITLE H=1.5 F=TRIPLEX C=PINK 'EWMA CONTROLLER SIMULATION'; TITLE3
H=0.7 F=SIMPLEX C=YELLOW
     "SPROC=&SPROC  THETA=&THETA  DB=&DB   CG=&CG"; TITLE4 H=0.7
F=SIMPLEX C=YELLOW
     "A1=&A1  B0=&B0  B1=&B1  B2=&B2  B3=&B3";
RUN;
```

Appendix 2

CUSUM SIMULATION PROGRAM

THIS PROGRAM SIMULATES AN CUSUM CONTROLLER WITH A
FIRST ORDER PLUS DEADTIME TRANSFER FUNCTION AND AN

IMA(0,1,1) DISTURBANCE PROCESS. IT RUNS ON VMS WITH SAS/BASIC AND SAS/GRAPH SOFTWARE, VERSION 5.16. A1=delta, and B0-B3 are parameters used to change omega and b. DB=K, SOUT=H, and SZERO provides a fast initial response feature. If SZERO=0, then there is no fast initial response, and if SZERO=1 the CUSUM initializes at H.

Different case studies are performed by changing the numbers in the following %LET statements. It is convenient to run the program from the full screen display (FSD) after removing this comment.;

```
OPTIONS DQUOTE;
%LET SPROC=1 ; %LET THETA=.50;
%LET A1=0 ; %LET B0=1  ; %LET B1=0  ; %LET B2=0 ; %LET B3=0 ;
%LET DB=.5  ; %LET SOUT=3.0  ; %LET SZERO=.5  ; %LET CG=0.5  ;
******************************************************
THIS PROGRAM SIMULATES A CUSUM CONTROLLER WITH A FIRST
ORDER PLUS DEADTIME TRANSFER FUNCTION AND AN IMA(0,1,1)
DISTURBANCE PROCESS.
*****************************************************;
DATA A;
DO TIME=1 TO 2000;T=0;
RETAIN SHTM1 SLTM1 NHM1 FHM1 NLM1 FLM1
ATM1 NTM1 XTM4 XTM3 XTM2 XTM1 YTM1 0;
************** DISTURBANCE  PROCESS  ***********;
AT=&SPROC*RANNOR(0);
NT=NTM1+AT-&THETA*ATM1;
************** TRANSFER  FUNCTION  ****************;
YT=&A1*YTM1+&B0*XTM1+&B1*XTM2+&B2*XTM3+&B3*XTM4;
ET=YT+NT;
PG=(&B0+&B1+&B2+&B3)/(1-&A1);
***************** UPPER  CUSUM  CONTROLLER  ***********;
DXT=0;
SHT=SHTM1+ET-&DB;
NH=NHM1+1;FH=FHM1;
IF SHT LT 0 THEN NH=0;
IF SHT LT 0 THEN FH=0;
IF SHT GT &SOUT AND FH=0 THEN DXT=-&CG*(&DB+SHT/NH)/PG;
IF SHT GT &SOUT AND FH=1 THEN DXT=-&CG*(&DB+(SHT-&SOUT*&SZERO)/NH)/PG;
IF SHT GT &SOUT THEN NH=0;
IF SHT GT &SOUT THEN FH=1;
IF SHT GT &SOUT THEN SHT=&SOUT*&SZERO;
IF SHT LT 0 THEN SHT=0;
***************** LOWER  CUSUM  CONTROLLER  ***********;
```

```
SLT=SLTM1-ET-&DB;
NL=NLM1+1;FL=FLM1;
IF SLT LT 0 THEN NL=0;
IF SLT LT 0 THEN FL=0;
IF SLT GT &SOUT AND FL=0 THEN DXT=+&CG*(&DB+SLT/NL)/PG;
IF SLT GT &SOUT AND FL=1 THEN DXT=+&CG*(&DB+(SLT-
&SOUT*&SZERO)/NL)/PG; IF SLT GT &SOUT THEN NL=0;
IF SLT GT &SOUT THEN FL=1;
IF SLT GT &SOUT THEN SLT=&SOUT*&SZERO;
IF SLT LT 0 THEN SLT=0;
******************** UPDATE VARIABLES **************;
IF DXT NE 0 THEN FREQ=1;ELSE FREQ=0;
XT=XTM1+DXT;
ATM1=AT;
NTM1=NT;
XTM4=XTM3;
XTM3=XTM2;
XTM2=XTM1;
XTM1=XT;
YTM1=YT;
SHTM1=SHT;
SLTM1=SLT;
NHM1=NH;
FHM1=FH;
NLM1=NL;
FLM1=FL;
********** DROP UNNEEDED VARIABLES FOR PRINTOUT *******;
DROP ATM1 NTM1 PG;
DROP XTM1 XTM2 XTM3 XTM4 YTM1;
DROP SHTM1 SLTM1 FHM1 NHM1 FLM1 NLM1;
OUTPUT;END;

****************** PROCEDURES ********************;
PROC MEANS MAXDEC=3;VAR ET DXT FREQ;
TITLE 'CUSUM CONTROLLER SIMULATION';
TITLE3
      "SPROC=&SPROC    THETA=&THETA";
TITLE4
      "A1=&A1  B0=&B0  B1=&B1   B2=&B2    B3=&B3";
TITLE5
      "SOUT=&SOUT  SZERO=&SZERO   DB=&DB   CG=&CG"; DATA B;SET
A;IF TIME LE 250;
PROC GPLOT;
PLOT (ET T)*TIME/OVERLAY
VAXIS=-16 TO 6 BY 2 HAXIS=0 TO 250 BY 50
```

```
VREF=-3,3 CV=RED CAXIS=YELLOW CTEXT=CYAN;
PLOT2 XT*TIME/
        VAXIS=-10 TO 30 BY 5 CAXIS=YELLOW CTEXT=ORANGE; SYMBOL1
I=JOIN C=CYAN ;
SYMBOL2 I=JOIN C=GREEN;
SYMBOL3 I=JOIN C=ORANGE;

TITLE H=1.5 F=TRIPLEX C=PINK 'CUSUM CONTROLLER SIMULATION'; TITLE3
H=.6 F=SIMPLEX C=YELLOW
        "SPROC=&SPROC    THETA=&THETA";
TITLE4 H=.6 F=SIMPLEX C=YELLOW
        "A1=&A1   B0=&B0   B1=&B1    B2=&B2    B3=&B3";
TITLE5 H=.6 F=SIMPLEX C=YELLOW
        "SOUT=&SOUT   SZERO=&SZERO   DB=&DB   CG=&CG";
RUN;
```

14
A BIVARIATE PROCESS CAPABILITY VECTOR

Norma Faris Hubele   H. Shahriari   C-S. Cheng*
Associate Director   Graduate Students
Statistical and Engineering Applications for Quality Laboratory
Industrial and Management Systems Engineering
Arizona State University
Tempe, AZ 85287-5906

Abstract

Capability defined in two dimensions should summarize simultaneously the performance on the two process variables vis-a-vis both sets of specification limits. In this paper, a three component capability vector based on a bivariate probability contour is proposed. The first component is analogous to the univariate C_p and provides information about the observed process variation relative to the region defined by both sets of specification limits. The second component, using a test statistic, summarize the relative location of the process and specification centers. The third component records the relative location of maximum and minimum values in the probability contours and the specification limits. Illustrations from actual manufacturing data are used to demonstrate the convenience of the capability vector for summarizing process performance.

Introduction

In modern manufacturing environments where complex processes require monitoring, the possibility of simultaneously

---

*Present affiliation: Associate Professor, Department of Industrial Engineering, Yuan-Tze Memorial College of Engineering, Taoyuan Shian, Taiwan, Republic of China.

monitoring and controlling more than one quality characteristic is rapidly becoming an important issue. When multiple quality characteristics are interdependent or correlated and subject to a system of chance causes, multivariate control methods are recommended. Alt [1985] provides a good review of the developments in the area of multivariate control schemes, originating with the work of Hotelling in 1947. Since 1985 numerous others have proposed additional control schemes, in particular, using extensions of the univariate CUSUM (scc Woodall and Ncube [1985], Alwan [1986]).

Critical to any statistical process control (SPC) system is a characterization of the capability of a process. Montgomery [1985] summarizes process capability as the uniformity of a process as measured by the variability of the process. This measurement is typically summarized in the form of a unit-less index, with simple rules for interpreting ranges of values of this index. As he points out, the results of process-capability studies may be used in product and process design, vendor sourcing, production or manufacturing planning and manufacturing.

The usual approach is to perform a process-capability analysis by studying a univariate quality characteristic of the product. The objective of this paper is to propose a composite measure of the capability of a process based on the analysis of two quality characteristics of the product produced. It is assumed that these characteristics are statistically dependent, and therefore, they should be monitored and controlled together. It follows that the capability of the process with respect to these characteristics should also be described together. A thorough review of the literature reveals that, to date, no work has been done on developing a bivariate capability measure.

Kane [1986] provides a very good discussion concerning univariate capability indices. As an introduction to the proposed bivariate capability index, we shall briefly outline the meaning of two very popular ones. In the univariate case, process capability ratio (C_p or PCR) is defined as the ratio of the allowable process spread to the actual process spread or

$$C_p = (USL - LSL) / 6\sigma \qquad (1)$$

The parameter s is the standard deviation of the process variable being measured. The value of 6s is sometimes called the manufacturing tolerance since it reflects the amount of tolerance that the process actually displays in manufacturing. The difference between USL and LSL, the upper and lower specification limits, is often called the design tolerance. When C_p is less than 1.0 then the process contains more variability than is permitted by design. Values of C_p greater than 1.0 imply that the process has less variability than allowed in the design. This is the preferred situation since small shifts in the process would minimize the risk of moving outside of the specification limits.

C_p does not provide any information about the relative **location** of the process and the specifications. Alternatively, an index called C_{pk} incorporates this information by using a scaled distance between the process mean and the closest specification limit. That is,

$$C_{pk} = \text{Min} \left[C_{pL}, C_{pU} \right] \tag{2}$$

where $C_{pL} = (\text{USL} - m)/ 3s$
$C_{pU} = (m - \text{LSL})/ 3s$
m is the process mean
s is the process standard deviation.

The C_{pk} index may take on any negative, positive or zero values. Similar to C_p, a value greater than 1.0 implies that the 6s limits fall completely within the specification limits. With this index it is clear when the process is operating within the specification limits.

In this paper a three component **process capability vector, CV**, based on a bivariate probability contour is proposed. The first component is analogous to the univariate C_p and provides information about the observed process variation relative to the region defined by both sets of specification limits. The second component, using a test statistic, summarizes the relative location of the process and specification centers. The third component records the relative location of maximum and minimum values in the probability contours and the specification limits.   Illustrations from actual manufacturing data demonstrate that the capability vector is a very convenient mechanism for summarizing   process performance.

Bivariate Process Capability Vector

As an extension to the distributional assumption for the univariate case, the process variables under discussion are assumed to be drawn from a bivariate normal distribution. This distribution may be uniquely defined by the parameters.

$$\mu = \begin{pmatrix} \mu_1 \\ \mu_2 \end{pmatrix} \; \Sigma = \begin{pmatrix} \sigma_{11} & \sigma_{12} \\ \sigma_{12} & \sigma_{22} \end{pmatrix} \tag{3}$$

It is well known that contours of constant density are ellipses   centered at m.   Further, the x values satisfying

$$(x-\mu)^T \, \Sigma^{-1} \, (x-\mu) \leq \chi_2^2(\alpha) \tag{4}$$

have probability 1-a.   That is, any bivariate observations drawn from this specified distribution will satisfy this inequality with probability 1 -a where χ_2^2 (a) is the upper (100 a)th percentile of a chi-square distribution with 2 degrees of freedom.   The region defined by (3) is called the 1-a probability contour.

Specification limits of process variables are generally defined as two discrete values, an upper limit and a lower limit, as discussed above.   In this paper, we are concerned with two process variables each with their own set of specification   limits. Let LSL_i and USL_i, where i=1,2, be the lower and upper specification limits for two quality characteristics.   Figure 1 illustrates the "specification rectangle" formed by these two sets of specification limits.   The vertices of this rectangle are (LSL_1, LSL_2), (USL_1, LSL_2) (LSL_1, USL_2), (USL_1, USL_2).   All bivariate observations of the two quality characteristics within or on this rectangle conform to specifications.

Recall that the univariate C_p index uses the scale of 6s to define capability.   This is based on the normality assumption

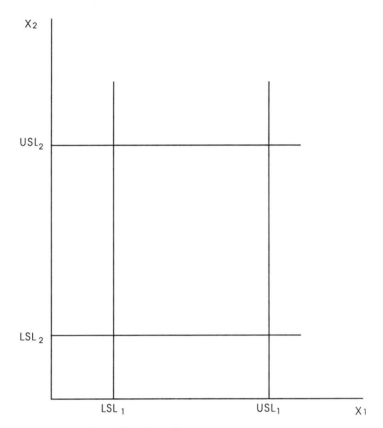

Figure 1. Specification Rectangle

that 99.73% of the distribution lines within the 6s limits.   If each variate of the bivariate observation were treated as a univariate observation, then the probability that both observations are within 3s of their respective means would be based on the $(0.9973 \times 0.9973) = 0.995$ probability contour of the bivariate distribution.

Bearing in mind that the probability contour of the underlying bivariate distribution is an ellipse, the objective of the formulation of the capability vector is to summarize the size and location of this contour representing the process relative to the specification rectangle.   Toward this end, define the smallest rectangle that can circumscribe the 0.995 probability contour of the process distribution as the "process rectangle." The sides of this rectangle are defined by the maximum and minimum values of the two process variables satisfying the equation of the ellipse. These values are determined by solving the system of equations of first derivatives of equation (4). As shown in Figure 2 the vertices of this rectangle are (LPR$_1$, LPR$_2$), (UPR$_1$, LPR$_2$), (LPR$_1$, UPR$_2$), (UPR$_1$, UPR$_2$).

The first component of the proposed capability vector is the ratio of the area of the specification rectangle and the area of the process rectangle. The meaning of this component is analogous to that of C$_p$; that is, a value greater than 1 implies that the process has smaller variation than allowed by the specification limits, a value less than 1 implies more variation.

In order to incorporate information about the relative location of the process and the specifications, a second index is

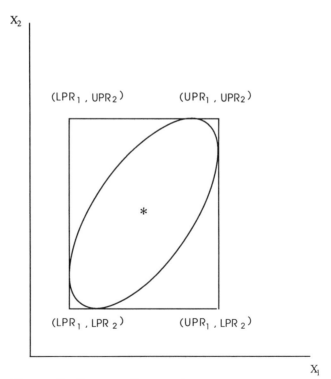

Figure 2. Process Rectangle

proposed. This component is defined as the significance level of the statistic computed from the difference between the center of the specification limits(m_0) and the estimated mean vector. As shown by Johnson and Wichern [1982], this statistic is computed as

$$T^2 = (\overline{x} - \mu_0)^T \left(\frac{S}{n}\right)^{-1} (\overline{x} - \mu_0) \tag{5}$$

where Hotelling's T^2 and follows a $[2(n-1)/(n-2)]F_{2, n-2}$ distribution. $F_{2, n-2}$ is obtained from the published tables. S is the estimate of S based on a sample of size n.

When the center of the observed distribution is "very close" to the center of the specification rectangle (assumed to be m_0), then this significance level of T^2 in (5) will be close to 1. When the significance level is close to 0, this implies that the centers are "not close."

The third component of the proposed vector provides additional information about the location of the sides of the process rectangle relative to the specification limits. Here, we are concerned with capturing the situation when one or more of the sides of the process rectangle fall outside the specification limits. Using the vertices of the specification and process rectangles, this component is defined as

$$\text{Max}\left\{1, \frac{|UPR_1 - LSL_1|}{USL_1 - LSL_1}, \frac{|LPR_1 - USL_1|}{USL_1 - LSL_1}, \frac{|UPR_2 - LSL_2|}{USL_2 - LSL_2}, \frac{|LPR_2 - USL_2|}{USL_2 - LSL_2}\right\} \tag{6}$$

When this index is equal to 1, then the entire process rectangle falls within or on the specification rectangle. When it is greater than one, then some or all of the process rectangle falls outside of the specification rectangle. The following illustrations will be useful for interpreting these two dichotomous situations. An interpretation of an exact value of this component is not recommended.

Illustrations

Table 1 and Figures 3, 4 and 5 provide illustrative examples of specification rectangles, process distributions and their

Table 1   Examples of Process Capability Vectors

FIGURE NUMBER	PARAMETER ESTIMATES					CAPABILITY VECTOR
	$\bar{x}_1$	$\bar{x}_2$	s_{11}	s_{22}	s_{12}	
3	2.15	3.77	1.54	2.72	1.77	[4.61, 0, 1.0)
4	1.36	-7.33	1.23	0.59	0.51	[17.80, 0, 1.0)
5	2.30	3.66	4.45	8.94	5.45	[0.96, 0.1, 1.34]

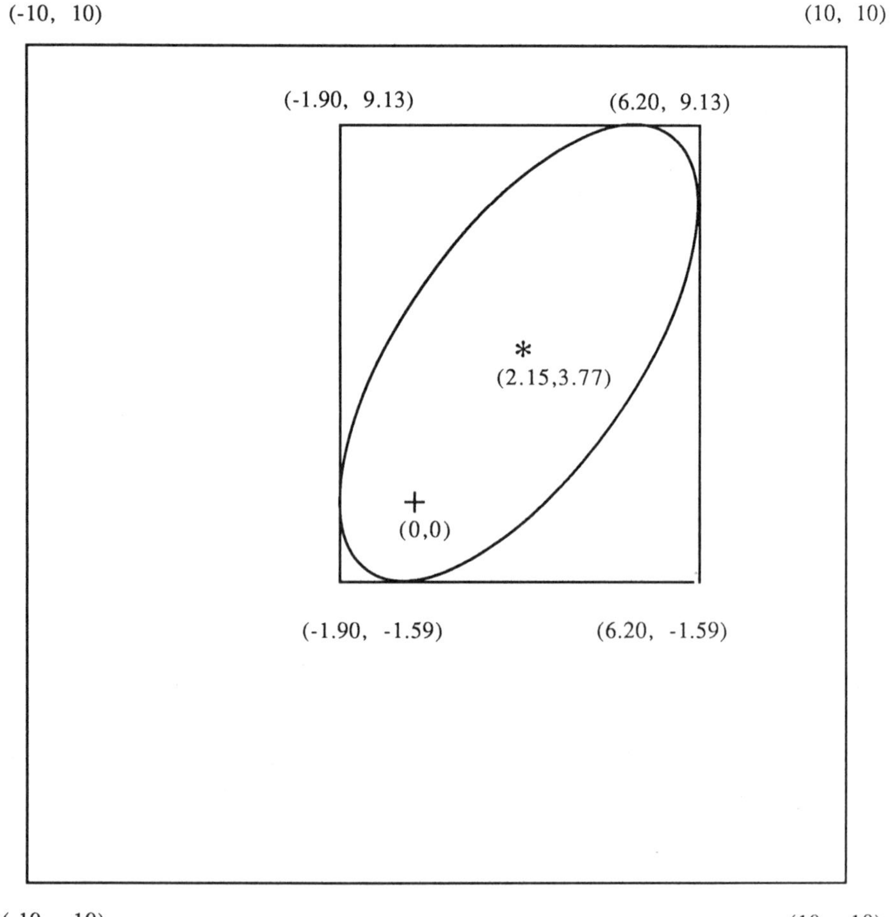

Figure 3. Process Capability Vector = [4.61, 0.0, 1.0]

associated capability vectors.   The first example reported in Table 1 and illustrated in Figure 3, has a specification rectangle (actually here it is a square) with area much larger than the process rectangle, hence the first component of the capability vector is 4.6.   With the relatively small values of the variances-covariances, the center of the process $\bar{X}$ = (2.15, 3.77) is significantly different from the nominal center m_0 = (0, 0) with

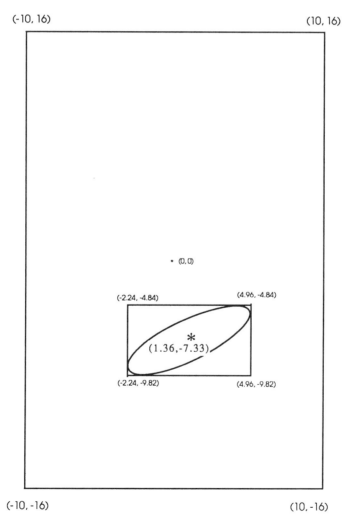

Figure 4. Process Capability Vector = [17.80, 0.0, 1.0]

a significance level approximately equal to 0. The third
component, 1.0, reflects the fact that the entire process
rectangle lies within the boundaries of the specification
rectangle.

In the second example shown in Table 1 and Figure 4, the
process rectangle is very small relative to the specification

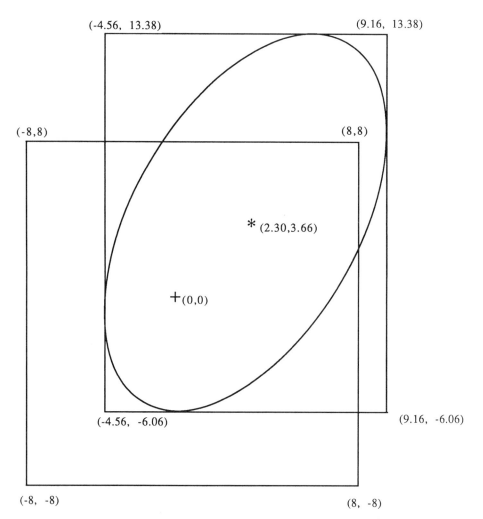

Figure 5. Process Capability Vector = [0.96, 0.1, 1.34]

rectangle resulting in a first component of the process capability vector equal to 17.8.   Again, the center of the process is significantly different from the nominal specifications and the entire process rectangle lies within the boundaries of the specification rectangle.

Figure 5 depicts the third example in which the area of the process rectangle rectangle exceeds that of the specification rectangle reflected in the first component of the vector being less than 1, i.e., 0.96.   With the relatively large values of variance-covariance, the centers of the specification and process are not significantly different at the .05 level.   The overlap of the process rectangle and the specification rectangle is captured in the third component with a value greater than 1.0, i.e., 1.34.

Conclusion

Process-capability analysis is an important part of statistical process control. If two variables are to be controlled together, then there needs to be a bivariate process capability measure. The vector proposed in this paper is a means of achieving this. It depends on observations drawn from a bivariate normal distribution.   The illustrations demonstrate the usefulness of this vector in summarizing the process.   Corrections to process problems, nevertheless, would still require examination of the marginal distributions.   Future research could include an extension of this vector to the general multivariate case.

References

Alt, F. (1985). "Multivariate Quality Control," "Encyclopedia of Statistical Sciences, 6 eds. S. Kotz and N. Johnson, Wiley, New York, pp. 110-122.

Alwan, L. (1986). "CUSUM Quality Control: Multivariate Applications," Communications in Statistics, Part A - Theory and Methods 15, pp. 3531-3543.

Johnson, R.; and Wichern, D. (1982). Applied Multivariate Statistical Analysis. Prentice-Hall, Englewood Cliffs, NJ.

Kane, V. (1986). "Process Capability Indices," Journal of Quality Technology 18, pp. 41-52.

Montgomery, D. (1985). Introduction to Statistical Quality Control. Wiley, New York, NY.

Woodall, W.; and Ncube, M. (1985). "Multivariate CUSM Quality - Control Procedures," Technometrics 27, pp. 285-292.

1 5
ECONOMIC DESIGN OF CONTROL CHARTS:
A REVIEW AND LITERATURE SURVEY
(1979-1989)

Lora Svoboda
Research Associate
Statistical and Engineering Applications for Quality Laboratory
College of Engineering and Applied Sciences
Arizona State University
Tempe, AZ 85287-5106

Introduction

This chapter is a review of current (1979-1989) literature regarding the economic design of control procedures. Work presented in this chapter follows the work that was surveyed and reviewed by Montgomery [1980]. This paper is restricted to economically-based procedures for product and process control purposes and does not include acceptance sampling.

The pioneering work in the area of economically-based control procedures was done by Girshick and Rubin [1952]. They were the first to consider the expected cost per unit time in quality control models. Duncan [1956] was the first to propose a model for the economic design of an X-bar control chart. Since this pioneering work and the development of Duncan's fully economic model, many other models have been developed. These models include the joint economic design of X-bar and R charts, economically-based CUSUM charts and economically-based fraction defective control charts.

Montgomery [1985] provides summaries and examples for many of these different types of control charts. His summaries also include the development of optimal production runs and sampling intervals based on production costs.

Literature concerning the economic design of control charts has been researched by several authors. Montgomery [1980] summarizes and reviews the literature in the area of economically-based control charts. His review discusses the pioneering work, development of Duncan's model, and models that were subsequently developed through the year 1979. Vance [1983] presents a bibliography of control chart techniques for the years 1970 to 1980. This bibliography includes, but is not limited to, economically-based control charts. v. Collani [1988] presents an updated bibliography of economic quality control procedures. This bibliography includes literature that has appeared since Montgomery's review; however, it does not summarize or review this literature. This chapter summarizes and reviews those articles which have appeared since 1979, many of which are included in v. Collani's bibliography. The literature is grouped into five main categories: variable control charts, attribute control charts, other models, weaknesses of economically-based control charts and plans for optimal process monitoring.

Variables Control Charts

The economic design of X-bar control charts was first introduced in 1956 by Duncan. In this early work, Duncan developed the economic model for an X-bar control chart but noted the need to develop models for joint X-bar and R charts. Subsequently, such models were considered and the results for both models (X-bar and joint X-bar and R) were reviewed by Montgomery [1980]. In this section the conclusions of Montgomery are outlined and updated with results obtained since 1979. Also, the recent important references for economically-based X-bar, joint X-bar and R and other control chart models are reviewed.

X-bar Control Charts.   Montgomery reviewed the work of Duncan and other authors whose work was based on Duncan's model and reached the following general conclusions regarding the economic design of X-bar Control charts: (1) the magnitude of the process shift largely determines the optimum sample size, (2) the interval between samples is mainly affected by the penalty cost for production in the out-of-control state and the changes in the average number of occurrences of assignable causes, (3) the width of the control limits is chiefly affected by the cost for investigating the assignable causes, (4) the design parameters are affected by the variation in the costs of sampling, (5) the optimum design is relatively insensitive to errors in estimating cost parameters, and (6) X-bar control charts should not be arbitrarily designed.

Since Montgomery's review, most of the work in the area of economically designed X-bar control charts has been that of enhancing or optimizing the model developed by Duncan. Several of the papers that have re-examined the assumptions upon which Duncan based his cost model are reviewed in this section. Generally, the research focuses on determining which assumptions are robust and the conditions under which the the assumptions are valid. Some of the assumptions that are researched include: (1) that the underlying process measurements are normally distributed, (2) that the time between the occurrence of assignable causes has an exponential distribution and is Markovian, (3) that the process continues to operate during the search for assignable causes, and (4) that the process cost parameters are known. Also reviewed in this section include is a paper which presents a computer program and two others which outline simplified approaches to Duncan's model.

Rahim [1985] considers the use of a non-normal probability distribution in the economic design of X-bar charts. He determines that a model based on a non-normal distribution is indeed useful for processes that cannot meet the assumption of normality. Along with the mean and variance, the parameters of skewness and kurtosis are used in the design. "Operator error" is also considered in the model.

An assumption often made in the economic design of X-bar control charts is that the time between the occurrence of assignable causes is exponentially distributed. In cases where there may be wear in the process, for example, this assumption may not be appropriate. Banerjee and Rahim [1988] examine this assumption using a Weibull distribution to represent process occurrences that may have an increasing hazard rate. They use the cost model proposed by Lorenzen and Vance (1986). Hu (1984) presents an abstract in which he investigates the economic design of an X-bar control chart based on a non-Poisson process shift. Hu assumes that there is a fixed sampling interval throughout the production run. Banerjee and Rahim compare their results with those found by Hu and determine that lower costs can be obtained using the non-uniform sampling scheme proposed in their paper.

In an earlier paper, Banerjee and Rahim (1987) present a model using a renewal equation approach to study the role of the probability model of assignable causes. Renewal equations are used to find the expected cycle length and expected cost per cycle. A Markovian model is first developed to introduce this approach. Then, a non-Markovian model is developed and applied to a Gamma distribution. The results show that renewal equations may be a more easily applied approach for Markovian models and that they may also be applied to non-Markovian models.

Other common assumptions made in the economic design of control charts are: (1) that the process is allowed to continue during the search for the assignable cause, or (2) that the process is shut down until the search is completed. Panagos, Heikes, and Montgomery [1985] compare the two assumptions in the economic design of X-bar control charts. The cost models for both assumptions are developed and their performance is tested using experimental design. The results show that stopping production during the search for an assignable cause is always more costly, unless there is a high penalty for operating out-of-control. These results also show that there is a cost penalty for wrongly specifying the process model.

Process cost parameters are necessary for the economic design of control charts. Parameters such as sampling costs, costs to investigate assignable causes and false alarms must be estimated. Pignatiello and Tsai [1988] investigate the optimal economic design of X-bar control charts when the process cost parameters are unknown. Using Duncan's cost model,

Pignatiello and Tsai consider the cost parameters with several different noise scenarios. A three level orthogonal array is used to test each noise scenario. Their results reveal that in a situation where cost parameters are difficult to specify, economic designs based on noise scenarios out-perform models that are designed assuming no noise.

Montgomery [1982] presents a computer program that generates the optimal economic design of an X-bar control chart using Duncan's model. The program is based on the assumptions that the process begins in control, that there is a known process shift due to a single assignable cause, that the time between assignable causes is exponentially distributed, and that the process continues to operate during the search for an assignable cause. The parameters which must be specified for the program include the cost of sampling, the cost of finding the assignable cause, the cost of investigating a false alarm, the penalty cost for operating in an out-of-control state, the type I error rate, the power of the chart, the time to sample and interpret the results, and the time to find the assignable cause. The program calculates and reports the sample size, control limit, sampling interval, alpha risk, power and cost for the optimal design. The program also generates several other designs which allow the user to choose among designs with near minimum costs. The program listing is also included in Montgomery's paper.

Montgomery and Storer [1986] explore the use of simplified model structures in the economic design of control charts. Duncan's cost model structure for the X-bar chart is used as the demonstration vehicle. Montgomery and Storer demonstrate that by eliminating all cost terms except those associated directly with sampling, passing defective units, and searching for assignable causes, a highly simplified model results. This simplified model is easier to optimize than the conventional model and provides solutions that are nearly cost optimal. The simplified model requires only half of the parameters that are required by the original Duncan model.

v. Collani [1986] presents another simplified method for designing economically-based X-bar control charts. He uses a procedure which allows for a reduction in parameters and simplifies the loss function.   The results show that v. Collani's simplified procedure compares favorably with the exact results found in Montgomery [1982].

Joint X-bar and R charts.   Since the work of Duncan, many authors have developed economic control chart models that include both the mean and variance of the process. Montgomery [1980] concludes that joint X-bar and R charts allow less frequent sampling than an X-bar chart, due to the increased power of the joint charts.   In this section, papers in which the process variance is considered in the models through the use of both range and variance charts are reviewed.   A paper presenting a computer program for the optimal economic design of joint X-bar and R charts is also discussed.

Saniga [1979] develops a model for economic design of joint X-bar and R control charts.   His model detects the shift in the variance of the process due to gradual deterioration in the system as well as the shift in the mean due to an assignable cause.   He also investigates the effects of three alternate process models (geometric, Poisson, and logarithmic).   Saniga's results reveal the importance of correctly identifying the process distribution for the optimal economic design of X-bar and R control charts.

Jones and Case [1981] present further research in the area of economic design of joint X-bar and R control charts.   Their model is based on four process stages of control for the mean and variance.   The results of Jones and Case disclose that there is a five percent decrease in cost when joint X-bar and R control charts are used rather than the X-bar model designed by Duncan.

Saniga [1987] proposes heuristic economic designs for X-bar and R control charts.   The heuristic approach is a simplification based on the assumptions that optimal designs are robust regarding cost parameters and that they normally have high power.   The heuristic design approach utilizes isodynes presented by Saniga [1984].   Isodynes are used to select the sample size based on the power of the chart.   Then, the sampling interval is chosen so that the cost function is minimized.   The heuristic economic design of X-bar and R control charts is a more easily applied approach and can be done without the use of a computer.

Saniga and Montgomery (1981) compare four process control methods based on their efficiency. The four models that are investigated are variables charts (X-bar and R-charts), attribute charts, a regular inspection policy and a "never-inspection" policy. The four procedures are compared using experimental design. The results show that for large, medium and small process shifts, X-bar and R-charts are the most economical. However, as the cost of sampling increases, the most economical procedures become attribute charts or a regular inspection policy. On no occasion did the "never-inspection" policy result in the lowest cost.

The development of economically-based X-bar and variance charts is presented in Rahim, Lashkari and Banerjee [1988]. Their cost model for the proposed design follows the unified approach developed by Lorenzen and Vance [1986]. The joint mean and variance control charts are compared to X-bar and R charts. Results of the comparison show that the X-bar and variance charts have a lower cost and greater power.

A computer program for the optimal economic design of joint X-bar and R charts is presented by Rahim [1989]. The program is based on the cost model proposed by Saniga and Montgomery. Rahim describes the program, its operation, and provides the program listing.

The work reviewed in this chapter, representing literature of the period 1979-1988, does not conflict with the conclusions drawn in Montgomery's review. Recent research has shown that it is important to understand the process model and to choose the economic design of variables control charts based on the process. Specifically, it is essential to specify whether the process continues to operate during the investigation for an assignable cause, what probability distribution will be used to model the process variables, whether or not the process distribution is assumed to be Markovian, and to determine whether there is drift in the process due to deterioration or wear. Also, methods have been developed to make the economic design of control charts an easier task. Several approximate and more easily applied approaches are presented and appear to perform favorably when compared to the exact results. Computer programs that generate the optimal economic design for variables control charts have been developed.

Attribute Control Charts

The first economic design of fraction defective control charts was developed by Ladany [1973]. Numerous other models and enhancements have appeared since this design. Montgomery reviews much of the published research that was completed before 1980. There have been several models developed since that time. Models that are discussed in this section include the use of curtailed sampling plans, attribute control charts for multiple assignable causes, an economic design for multi-stage production systems and a simulation method for designing attribute control charts.

Williams, Looney, and Peters [1985] introduce the use of curtailed sampling plans in the economic design of np control charts. In a curtailed sampling plan, sampling is ceased once enough information has been gathered to make a decision about the process. The proposed model is an adaptation of earlier economic np control chart designs. The theoretical and empirical results reveal that although the curtailed sampling plan is only a modest improvement, it is never more costly than the fixed sampling plan.

The economic design of fraction defective control charts for multiple assignable causes is presented by Gibra (1981). The proposed model for the optimal design of np control charts considers the occurrence of several assignable causes. The model is developed for two process models. The effects of the cost parameters and risks on the optimal design are summarized in the paper. Gibra also proposes that a single cause model may be designed to match the multiple cause model. His results show that this more easily applied, "matched" single assignable cause model can be accurately used in a situation requiring a multiple cause model.

Peters and Williams [1987] propose a method for determining np chart parameters for each stage in a multiple stage process. Previous work in the area of multi-stage systems assumed that the proportion defective was constant for each stage. Their model allows the proportion defective to increase when there is an assignable cause at each stage.

Sculli and Woo [1985] present a simulation program to determine the optimal design of np control charts. The program is based on the relaxation of some of the assumptions found in literature regarding the development of economically designed control charts. It is also proposed that the simulation may be extended to determine optimal economic designs for X-bar, c, u and CUSUM charts.

The results of the more recent efforts in the area of economic design of attribute control charts have been the development of simplified models and the use of computers. Work aimed at making the economic design of control charts a more easily applied tool will accelerate its implementation into industry. Other research has provided an enhancement to the optimal economic design of attribute control charts for multi-stage systems.

Economic Design of Other Control Charts

Although the major effort in the area of economic design of control charts is based on Duncan's model, other cost models and designs have been developed. These include an approach based on a more general model, an acceptance control chart, a model based on the minimax principle and a design for multiple control limits.

Lorenzen and Vance [1986] present a unified approach for economically designing control charts. They observe that past work in the area of economic design of control charts has been fragmented. Many different models have been developed. Therefore, they develop a general model that may be used with control charts based on any statistic. Their model varies sample size, sampling interval and control limits to minimize an hourly cost function.

Tagaras and Lee [1988] propose an economically designed control chart for different assignable causes. They point out that a single pair of control limits may not be adequate when multiple assignable causes are present. Multiple assignable causes may require different degrees of corrective action. A model with two pairs of control limits is developed and compared to a single-cause model in which there is only one search and one reaction procedure. The results show that the single-cause, single-response models are more costly.

Bergman [1987] presents an improved acceptance control chart. An acceptance control chart is designed to to determine whether a process is acceptable by applying significance tests to successive samples from the process. Bergman improves upon the acceptance chart by applying cost model elements. His results show that the risk and power of the acceptance control chart remain the same, but the cost is decreased.

Another model, presented by Arnold [1987], is based on the pessimistic minimax-principle. This model is compared to previous work by both v. Collani and Arnold in which an optimistic model is considered. Both models conclude that the simple X-bar chart should be preferred over other, more complicated, charts.

Several of the models in this section disclose that the development of economically-based control charts is often concentrated toward specific situations. Lorenzen and Vance [1986], however, state the need for a more general model that may be applied to any control chart. This approach is useful for those who cannot find an approach that matches their process model or the statistic used to describe their process variables.

Weaknesses in the Economic Design of Control Charts

Although the economic design of control charts has been shown to have some advantages over statistically designed control charts, there are some disadvantages. Woodall [1986] points out some of the weaknesses including: (1) that economic control charts can have poor statistical performance, (2) there is often a high rate of false alarms, (3) the economic method is not effective in detecting small shifts, (4) the economic design is not consistent with Deming's philosophy, and (5) that the economic design has not changed greatly since it was developed by

Duncan. Woodall [1987] expands on these statements. He gives two examples, one for the CUSUM and another for the np chart. He adds that it may be more effective to use Taguchi's loss function or to economically determine the smallest shift to be detected.

Additionally, several other weaknesses are apparent in economically designed control charts. Although there has been recent research to simplify the economic design of control charts and to utilize computer programs, the control charts do not appear to be easily applied. Also, the necessary process cost parameters may not always be available.

Plans for Optimal Monitoring and Production Intervals

Determining optimal production runs and inspection plans are important factors in Quality Control. However, before the optimal production run and inspection plan can be determined, the critical product and process variables must be identified and their optimal level determined. Once this is done, the process can be monitored based on the optimal production intervals and inspection plans. This will ensure that the highest quality has been identified and is constantly being maintained.

Several models that are used to calculate the optimal production run have been developed and are presented in this section. These papers are based on models that balance the cost of adjusting the system and the cost of defective parts. Adjustments to the system are necessary when wear or deterioration inherent in the process causes shifts in the process mean and variance. Also reviewed in this section, are papers that discuss an inspection plan based on the inspection precision level, a comparison of periodic and continuous inspections policies and choosing zero or 100 percent inspection based on the Deming inspection criterion.

Bisgaard, Hunter and Pallesen [1984] present a paper that discusses determining the optimal target value for the process before applying quality control. They prove that the optimal target value is dependent on the economic conditions and the variation in the process. Arcelus and Banerjee (1985) extend the model developed by Bisgaard, Hunter and Pallesen to consider a process where there is a shift in the mean after the optimal value has been determined. Using this model, once the optimal target value is determined, the optimal production run must also be determined to maximize the the unit profit.

Taguchi, Elsayed, and Hsiang [1989] present an on-line process control system based on Taguchi's loss function. The model is based on the loss due to the costs to measure and adjust the system and the costs associated with reworking or scrapping units outside the tolerance range. The model assumes that once the process is out-of-specification, it continues in that state until an adjustment is made. The model is used to determine the optimal inspection interval, the control limits and the adjustment interval. These parameters are all optimized to minimize the total loss. Models for both variables and attribute characteristics are developed. A variables on-line control model is presented in Taguchi [1986].

Arcelus, Banerjee, and Chandra [1982] present a model to determine the optimal production run. They consider the situation where there is a shift in the mean and variance due to wear in cutting tools. Their objective is to determine the number of acceptable parts that can be produced while minimizing the total cost of production. They offer solutions to both discrete and continuous cases. This model considers only the profit gained from parts that are acceptable. Arcelus and Banerjee (1987) generalize their earlier model to incorpoate the rewards that may be gained from the use of defective parts. Rewards include scrap sales for under-sized parts and rework or scrap sales for over-sized parts. The objective of this model is to maximize the profit from each unit that is produced.

Rahim and Lashkari [1985] present a model that uses double specification limits for a process containing shifts in both the mean and variance. This model is an extension of an earlier model developed by Rahim and Lashkari where they only a shift in the mean was considered. They provide three examples: (1) a process having shifts in mean and variance, (2) a process only affected by drift, and (3) a negative shift in the process mean due to an assignable cause and drift in the process. They conclude that magnitude and direction of the shift and the drift are the main factors in determining the optimal production run.

McDowell [1987] proposes a model for the economic design of a sampling control strategy for a class of processes that are easily and inexpensively adjusted. Schmidt and Pfeifer [1989] develop a model showing that there is a linear relationship between decreasing the process variation and lowering costs. They propose that this model may be used to prioritize and select process quality improvement projects.

Tang and Schneider [1988] develop a model that optimizes the precision level of the sampling technique for the process. The model balances the cost of obtaining a high precision level and the cost benefits from this more precise level. The model was developed for a 100% inspection plan. Rosenblatt and Lee [1986] compare the use of continuous and periodic inspection policies in a deteriorating production system. They provide results that are useful in determining which policy will be the most effective based on process assumptions and costs.

Papadakis [1985] presents Deming's inspection criterion. Deming points out that acceptance sampling may not be the lowest total cost option. Deming assumes that the fraction defective is fixed rather than variable. He determines the ratio of the downstream defective to the inspection costs and multiplies this with the fraction defective to decide upon zero or 100% inspection. Sengupta [1982] suggests that, by knowing when to inspect the system, the expected cost incurred due to the occurrence of defectives can be minimized.

General conclusions regarding optimal production runs and inspection plans highlight the importance of: (1) determining the target mean for the the critical variables, (2) designing the plan to monitor the process including process drift and shift due to assignable causes, (3) including all cost parameters in the design, and (4) understanding the magnitude and direction of the shift and drift. With the need to become high quality, low cost producers, the research in this area will help many industries. Many of the methods are straight forward and can be easily applied in industry. For example, the method proposed by Taguchi et al. is not complicated and is not difficult to use.

Literature Not Reviewed

Listed after the references at the end of the chapter are citations which were unavailable for review.

Conclusion

While the optimal economic design of control charts may offer reductions in cost, they are more difficult to apply than "risk-based" control charts. However, due to the recent research the application of economically-based control charts through the use of approximations and computer programs has been simplified.

It is necessary to determine the optimal production run and inspection interval to maintain high quality products and to minimize production costs. Once the optimal levels for the product or process have been identified, it is important to balance the costs that are necessary in order to maintain these levels with the costs of defective parts. Many of the papers reviewed in this chapter offer models that can be used to optimally monitor a process.

References

Arcelus, F. J., and Banerjee, P. K. (1985). "Selection of the Most Economical Production Plan in a Tool-Wear Process." Technometrics 27, pp. 433-437.

Arcelus , F. J., and Banerjee, P. K. (1987). "Optimal Production Plan in a Tool Wear Process with Rewards for Acceptable, Undersized and Oversized Parts." Engineering Costs and Production Economics 11, pp. 13-19.

Arcelus, F. J., Banerjee, P. K., and Chandra, R. (1982). "The Optimal Production Run for a Normally Distributed Quality Characteristic Exhibiting Non-Negative Shifts in the Process Mean Variance. IIE Transactions 14, pp. 90-98.

Arnold, B. F. (1987). "Minimax-Principle and Control Chart Design." Frontiers in Statistical Quality Control 3. Lenz H.-J., Wetherill, G.B., and P.-Th. Wilrich (Eds.). Physica-Verlag, Heidelberg, pp. 229-241.

Banerjee, P. K., and Rahim, M. A. (1987). "The Economic Design of Control Charts: A Renewal Theory Approach." Engineering Optimization 12, pp. 63-73.

Banerjee, P. K., and Rahim , M. A. (1988). "Economic Design of x-bar-Control Charts Under Weibull Shock Models." Technometrics 30, pp. 407-414.

Bergman, B. (1987). "On an Improved Acceptance Control Chart." Frontier in Statistical   Quality Control 3. Lenz H.-J., Wetherill, G.B., and P.-Th. Wilrich (Eds.). Physica-Verlag, Heidelberg, pp. 154-162.

Bisgaard, S., Hunter, W. G., and PALLESEN, L. (1984). "Economic Selection of Quality of Manufactured Product." 26, pp. 9-18.

v. Collani, E. (1986). "A Simple Procedure to Determine the Economic Design of an X-bar Control Chart." Journal of Quality Technology 18, pp. 145-151.

v. Collani , E. (1988). "An Updated Bibliography of Economic Quality Control Procedures." Economic Quality Control 3, pp. 48-62.

Duncan, A. J. (1956). "The Economic Design of X-bar-Charts Used to Maintain Current Control of a Process." Journal of the American Statistical Association 51, pp. 228-242.

Gibra, I. N. (1981). "Economic Design of Attribute Control Charts for Multiple Assignable Causes." Journal of Quality Technology 13, pp. 93-99.

Girshick, M. A., and RUBIN, H. (1952). "A Bayes' Approach to a Quality Control Model." Annals of Mathematical Statistics 23, pp. 114-125.

Hu, P. W. (1984). "Economic Design of an X-bar Control Chart

Under Non-Poisson Process Shift." Abstract TIMS/ORSA Joint National Meeting, San Francisco. May 14-16, pp. 87.

Jones, L. L., and Case, K. E. (1981). "Economic Design of a Joint X-bar and R-Control Chart." AIIE Transactions 13, pp. 182-195.

Ladany, S. P., (1973). "Optimal Use of Control Charts for Controlling Current Production." Management Science 19, pp. 763-772.

Lorenzen, T. J., and Vance L. C. (1986). "The Economic Design of Control Charts: A Unified Approach." Technometrics 28, pp. 3-10.

McDowell, E. D. (1987). "The Economic Design of Sampling Control Strategies for a Class of Industrial Processes." IIE Transactions 19, pp. 289-295.

Montgomery, D. C. (1980). "The Economic Design of Control Charts: A Review and Literature Survey." Journal of Quality Technology 12, pp. 75-87.

Montgomery, D. C. (1982). "Economic Design of an X-bar Control Chart." Journal of Quality Technology 14, pp. 40-43.

Montgomery, D. C. (1985). Introduction to Statistical Quality Control. Wiley & Sons, Inc., New York.

Montgomery, D. C., and Storer, R. H. (1986). "Economic Models and Process Quality Control." Quality and Reliability Engineering International 2.

Panagos, M. R., Hiikes, R. G., and Montgomery, D. C. (1985). "Economic Design of X-Control Charts for Two Manufacturing Process Models." Naval Research Logistics Quarterly 32, pp. 631-646.

Papadakis, E. P. (1985). "The Deming Inspection Criterion for Choosing Zero or 100 Percent Inspection." Journal of Quality Technology 17, pp. 121-127.

Peters, M. H., and Williams, W. W. (1987). "Economic Design of Quality Monitoring Efforts for Multi-Stage Production Systems." IIE Transactions 19, pp. 81-87.

Pignatiello, J. J. Jr, and TSAI, A. (1988). "Optimal Economic Design of X-bar Control Charts When Cost Model Parameters are Not Precisely Known." IIE Transactions 20, pp. 103-110.

Rahim, M. A. (1985). "Economic Model of X-bar-Charts Under Non-Normality and Measurement Errors." Computers and Operations Research 12, pp. 291-299.

Rahim, M. A. (1989). "Determination of Optimal Design Parameters of Joint X-bar and R Charts." Journal of Quality Technology 21, pp. 65-70.

Rahim, M. A., and Lakshari, R. S. (1984). "Optimal Decision Rules for Determining the Length of the Production Run." Computer and Industrial Engineering 9, pp. 195-202.

Rahim, M. A., Lakshari,R. S., and Banerjee, P. K. (1988). "Joint Economic Design of Mean and Variance Control Charts." Engineering Optimization 14, pp. 65-78.

Rosenblatt, M. J., and Lee, H. L. (1986). "A Comparative Study of Continuous and Periodic Inspection Policies in Deteriorating Production Systems." IIE Transactions 18, pp. 2-9.

Saniga, E. M. (1979). "Joint Economic Design of X-bar and R Control Charts with Alternate Process Models." AIIE Transactions 11, pp. 254-260.

Saniga, E. M. (1984). "Isodynes for X-bar and R Control Charts." Frontiers in Statistical Quality Control 2. Lenz H.-J., Wetherill, G.B., and P.-Th. Wilrich (Eds.). Physica-Verlag, Wurzburg, pp. 268-273.

Saniga, E. M. (1987). "Heuristic Economical Design for X-bar and R Control Charts." Frontiers in Statistical Quality Control 3. Lenz H.-J., Wetherill, G.B., and P.-Th. Wilrich (Eds.). Physica-Verlag, Heidelberg, pp. 220-228.

Saniga, E. M., and Montgomery, D. C. (1981). "Economically Quality Control Policies for a Single Cause System." AIIE Transactions 13, pp. 258-264.

Schmidt, R. L., and Pfeifer, P. E. (1989). "An Economic Evaluation of Improvements in Process Capability for a Single-Level Canning Problem." Journal of Quality Technology 21, pp. 16-19.

Sculli, D., and Woo, K. M. (1985). "Designing Economic np Control Charts: Programmed Simulation Approach." Computers in Industry 6, pp. 185-194.

Sengupta, B. (1982). "An Exponential Riddle." Journal of Applied Probability 19, pp. 737-740.

Tagaras, G., and Lee, H. L. (1988). "Economic Design of Control Charts with Different Control Limits for Different Assignable Causes." Management Science 34, pp. 1347-1366.

Taguchi, G., (1986). Introduction to Quality Engineering. Asian Productivity Organization, Tokyo.

Taguchi , G., Elsayed , E. A., and Hsiang, T. (1989). Quality Engineering in Production Systems. McGraw-Hill, New York.

Tang, K., and Schneider, H. (1988). "Selection of the Optimal Inspection Precision Level for a Complete Inspection Plan." Journal of Quality Technology 20, pp. 153-156.

Vance, L. C. (1983). "Bibliography of Quality Control Chart Techniques." Journal of Quality Technology 15, pp. 59-62.

Williams, W. W., Looney, S. W., and Peters, M. H. (1985). "Use of Curtailed Sampling Plans in the Economic Design of np-Control Charts." Technometrics 27, pp. 57-63.

Woodall, W. H. (1986). "Weakness of the Economic Design of Control Charts." Technometrics 28, pp. 408-409.

Woodall, W. H. (1987). "Conflicts Between Deming's Philosophy and the Economic Design of Control Charts." Frontiers in

Statistical Quality Control 3. Lenz H.-J., Wetherill, B. G., and Wilrich P.-Th. (Eds.). Physica-Verlag, Heidelberg, pp. 242-248.

Additional   References

Arnold, B. F. (1985). "Approximately Optimum Two-Stage Sampling Plans." Metrika 32, pp. 293-313.

Arnold, B. F. (1986). "Procedure to Determine Optimum Two-Stage Sampling Plans by Attributes." Metrika 33, pp. 93-109.

Arnold, B. F., and v. Collani, E. (1989). "On the Robustness of X-bar Charts." Statistics (in press).

Beighter, C. S., Phillips, D. T., and Wilde, D. J. (1979). Foundations of Optimization. Prentice-Hall Inc., Englewood Cliffs, NJ.

Chiu, W. K., and Leung, M. P. Y. (1980). "A New Bayesian Approach t0 Quality Control Charts. " Metrika 27, pp. 243-253.

v. Collani, E. (1981). "On the Choice of Optimal Sampling Intervals to Maintain Current Control of a Process." Frontier in Statistical Quality Control. Lenz H. J., Wetherill, B. G., and Wilrich P.-Th. (Eds.). Physica-Verlag, Wurzburg, pp. 242-248.

v. Collani, E. (1987). "Economic Control of Continuously Monitored Production Processes." Rep. Stat. Appl. Res., JUSE 34, pp. 1-18.

Ohta, H. (1980). "GERT: Analysis of the Economic Design of Control charts." IAPQR Transactions 5, pp. 57-65.

Rahim, M. A., and Lashkari, R. S. (1981). "An Economic Design of X-bar Charts with Warning Limits to Control Non-Normal Process Means." Proceedings Computer Science and Statistics 13, pp. 371-374.

Rahim, M. A., and Lashkari, R. S. (1983). "A comparison of Economic Designs of Non-Normal Control Charts." Proceedings Computer Science and Statistics 15, pp. 368-372.

Rahim, M. A., and Lashkari, R. S. (1983). "Optimal Production Run for a Process Subject to Drift, Shift, and Changes in Variance." Proceedings Computer Science and Statistics 14, pp. 291-294.

Raouf, A., Jain, J. K., and Sathe, P. T. (1981). "A Model for Determining the Optimal Number of Repeated Inspections for Complex Units to Minimize the Total Cost." Rep. Stat. Appl. Res., JUSE 28, pp. 1-11.

Rinne, H. (1981). "Cost Minimal Process Control." Frontiers in Statistical Quality Control. Lenz H.-J., Wetherill, B. G., and Wilrich P.-Th. (Eds.). Physica-Verlag, Wurzburg, pp. 227-241.

Sculli, D., and Woo, K. M. (1982). "Designing np-Control Charts." Omega, International Journal of Management Science 10, pp. 679-687.

Shahani, A. K. (1981). "Choice of Process Inspection Scheme: Some Basic Consideration." Frontiers in Statistical Quality Control. Lenz H.-J., Wetherill, B. G., and Wilrich P.-Th. (Eds.). Physica-Verlag, Wurzburg, pp. 275-294.

Sultan, T. I. (1981). "A Model for Economic Design of X-bar Control Charts with Several Assignable Causes." ASQC Technical Conference, pp. 408-411.

1 6
ON-LINE QUALITY CONTROL FOR THE FACTORY OF THE 1990'S AND BEYOND

Frederick W. Faltin and William T. Tucker
Management Science and Statistics Program
GE Corporate R & D Center
Schenectady, NY   12301

Introduction

The business shake-out of the 1980's has clearly demonstrated that American industry, if it is to survive, must eradicate the cost and quality disadvantages from which it has been competing in world markets. Harsh competitive realities dictate that higher workforce productivity and greater uniformity of product must become the rule rather than the exception.

The advent of automated manufacturing has provided industry a tool with which to confront this challenge. Probing the limitations of current process technology has become a matter of economic necessity. As might have been expected, the initial steps in this new direction have at times been faltering ones. Even setbacks, however, have led not to abandonment of the concept, but to constructive redirection of its application. Reason therefore suggests that, far from being a flash in the industrial pan, the present nascent state of automated manufacturing represents Western society's embarkation upon a new and abiding aspect of the Industrial Revolution.

Further supporting this view is the evident likelihood that developing technologies will continue to open new vistas throughout the foreseeable future. Fueled by the availability of microprocessors of vastly increased power and significantly decreased cost, advances in computer-controlled manufacturing equipment and peripherals are moving from development to commercialization. Ongoing work in sensory, vision, and knowledge-based systems will imbue the next generation of programmable manufacturing cells with enhanced capabilities and greater flexibility. Research in data storage and retrieval, data communications, and distributed systems, has and will continue to pay dividends in the integration of these units into broadly functional manufacturing systems.

Characteristic of the resulting automated factory environment is its emphasis on computing resources. These will typically range from dedicated special-purpose microprocessors at the extreme, through advanced microcomputers employed, e.g., as manufacturing cell controllers, to super-minis or mainframes as factory "hosts" at the upper end of the computing spectrum. Professional workstations will provide plant engineers and management with system access for control, monitoring, and analysis functions, while "dumb" terminals or lesser microcomputers support the diagnostic, maintenance, and repair efforts of factory technicians. This diverse array of processors will be united by an extensive data network, allowing distribution of computing and data collection/storage/retrieval tasks among the various components, according to their functions and capabilities.

Into this complex scenario must be integrated another force which has been gaining momentum as a consequence of international competitive pressures--the use of statistical techniques for improving quality and productivity. The magnitude of the challenge this poses may not, at first, be readily apparent. But even a moment's reflection convinces one that rote application of time-honored SOC practices developed for the production processes of yesteryear will not suffice. The picture of line operators plotting Shewhart charts by pencil is woefully obsolete in an environment where automated instrumentation can capture literally thousands of data items per hour on dozens, perhaps hundreds, of process and product variables. Simply replacing the operator with an engineer, and the pencil with hardware and software which automate what was previously done by hand, does not represent much of an improvement. Such uninspired attempts could not possibly keep pace with the potential even of existing systems; they could not begin to exploit the immense resources that the automated factory will possess.

It is tempting to assume that, in a futuristic manufacturing environment, fears of suboptimal quality and productivity will be as obsolete as the methods of production which once gave rise to them. Surely, reliable hardware and intelligent software will provide the ultimate in uniformity of product, and eliminate human flaws which are impediments to productivity! And with the implementation of adaptive controls, incipient departures from nominal performance will be corrected automatically, will they not?

These gross oversimplifications are all the more deceptive because they are true--up to a point. They are, after all, among the reasons which continue to motivate the installation of automated production processes. Moreover, successful applications attest to the fact that these benefits can, in fact, be attained. But to stop there would be to ignore the lessons of history, the dictates of common sense, and the reality of attempts which have fallen short. Lest we be too easily seduced, let us reflect further upon the challenges that automated production facilities must face.

The history of technology demonstrates that advances which exceed the needs of the present are soon barely adequate to those of the future. The reasons for this are many, but for our purposes it suffices to remark that design and production are complementary phases of a never-ending cycle. New manufacturing capabilities quickly lead to the development of products or processes which exploit those capabilities to (or PAST) the current state of the art. Economic expectations are adjusted accordingly. Yesterday's wonders are taken for granted today, and new projects are envisioned which require the latest advances for their very conception.

Automated manufacturing is no exception to this pattern. Where conventional production is labor-intensive, automated

factories are capital-intensive. Especially in a time of lingering
high real interest rates, investments in high-tech processes are
not likely to be regarded as attractive if they promise "only"
reduced labor costs (indeed, not all of the industries being
automated are labor intensive even at present. If they are to
succeed, automated factories must not just turn out products
more cheaply than conventional facilities--they must do it
BETTER, and they must do it FASTER. In machining applications,
automated tooling is expected today to turn, to within several
ten-thousandths of an inch, parts which were manually
produced to tolerances of thousandths only a few years ago. In
automobiles, stronger welds result in less welding, and
correspondingly higher throughput. In the chemicals industry,
improved process control means more consistent intermediates,
higher yields, and better end use performance. In all
industries, greater achievements breed still higher expectations.

Once one realizes that technological advances alone cannot
assure excess process capability, the continuing need for
statistical techniques to monitor and optimize production
processes becomes evident. We have already argued, however,
that simplistic automation of current methodology will not
suffice. In the sections which follow, we address these issues,
discussing the expanded role which statistical methods should
be expected to play in the new production environment, and
proposing a current-technology approach for use in first
generation automated plants. We conclude by sketching the
form of a next-generation system, and proposing areas where
research is needed to support its development.

SPC Systems in an Automated Factory Environment— Their Role and Goals

The principal changes that we foresee in statistical quality
systems for automated environments are in: (1) the choice of
what process characteristics are to be observed by statistical
monitoring techniques, such as control charts; (2) a much
expanded role for general statistical analyses, especially
exploratory data analysis; and, (3) the manner in which these
functions will b implemented in the plant.

This section will discuss the course we expect each of these
activities to follow, and the rationale behind our speculations.

Subsequent sections will propose how these ends might be achieved, first in the short to intermediate term, and, finally, in a longer-term view.

<u>Statistical Process Monitoring</u>. Historically, manually-kept control charts were used as devices for detecting process changes. An out of control signal triggered determination (necessarily by humans) of the cause of the change, which was followed in turn by corrective action (to reverse detrimental changes, or make beneficial ones permanent). The variables monitored were usually final product properties, often the same ones used to define whether or not the product conformed to specifications. In recent years, a new quality philosophy has given preference to monitoring of process parameters, in-process product properties, and raw materials characteristics. Though better reflecting the aim of building in quality "up front", the purpose has remained that of signaling process changes as a basis for human investigation and response.

One of the principal changes we anticipate for statistical monitoring in automated environmentsis is its integration with process control algorithms and software. As will become evident, we expect this to lead to a corresponding re-emphasis in the quantities monitored, to evolution of the role of SPC, and hopefully, to a long-overdue reassessment of the meaning of "out of control" signals.

In our view, statistical monitoring of final product properties will probably continue to decline in both conventional and automated facilities (although some is likely to continue in order to verify product compliance, provide customer documentation, and assess overall process capability). In hybrid plants, which operate automated equipment under manual control (and which are likely to remain common for quite a while in some industries before gradually going the way of the dinosaur) the principal reason for this re-orientation will be as it is now--to prevent bad product, rather than detecting it. But as more wholly automated factories operating under computerized adaptive controls emerge, the rationale for downplaying final product QC will be even more fundamental. In such facilities, the only nonrandom variation observable in the final product will be that due to system failures in the control apparatus, or to process changes of such a magnitude that the control scheme was unable to compensate. Both of these constitute essentially catastrophic failures which the plant quality system must be made capable of detecting at a much earlier stage. IN AUTOMATED FACILITIES, THEREFORE, PLACING STRESS ON UP FRONT QUALITY CONTROL IS NOT MERELY A PHILOSOPHY, BUT A TECHNICAL NECESSITY.

Variables to be Monitored. To understand which variables the quality system should be monitoring, we need to consider, in general, how an automated plant functions. Figure 1 presents a schematic of a plant control system. In this (admittedly oversimplified) view, the plant controller operates on some product stage by inputting data from the process to a model, which outputs instructions for adjusting process mechanisms to maintain or restore proper process function. For such a scheme to operate successfully, the key elements (assuming the absence of communication or computational malfunctions not germane to the present discussion) are therefore: (1) the material(s) being processed; (2) the data reaching the controller (which is used as input to its calculations); (3) the adequacy of the controller's process model; (4) the proper response of process mechanisms to the controller's instructions. These, then, are the critical system components which must be monitored by the statistical quality system. The status and/or function of each component can be verified on the basis of information observable during the operation of the process. Each component therefore defines a class of observables by which it may be characterized. Identifying these variables, and capturing and maintaining appropriate data on them, are essential activities in the design and implementation of any comprehensive quality system.

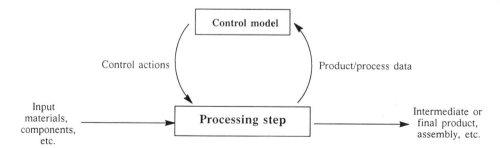

Figure 1. Simplified Plant Control System

The number, relative importance, and actual identities of the variables to be monitored will, of course, be highly industry-, process-, and installation-specific. Often there will be some overlap among categories. A few typical examples will serve to clarify our use of the terms employed above, and to provide a basis for further discussion of each.

Materials Data. In machining applications, the material being processed might be raw castings for forgings, workpieces, which have undergone intermediate turning, milling, or other steps, on through to the finished part. In the chemicals industry, it may be anything from the raw material stream(s), through any of a number of intermediate process products, to the final product itself. In autos, electronics, or major appliances, this could include incoming raw materials or assemblies, the status of locally produced subassemblies at any stage of their manufacture, and so forth, on to the final product. In consonance with current practice and or comments above, we view information of this sort as being of greater value, the further upstream (i.e., the earlier) in the process its origin (subject to an exception to be detailed later). Even so, our experience suggests that other categories of process information will often be of greater use in statistical process monitoring, for reasons which will become evident as we proceed.

Controller Input Data. Among the most valuable sources of data for study via SPC in automated factories is likely to be the data input to the control scheme (we stress here that we are referring to INPUT data). In machining applications, this is likely to consist of dimensional data from intermediate processing steps, and possible physical data on machine function, such as tool bit pressures, power consumption, etc. In the chemical industry, these might be reactor parameters (pressure, temperature, impeller rpm's), on-line analyzer data on various materials (viscosities, product mix, impurity levels), etc. In robotics applications, this could include sensory input from vision systems, and so forth.

At present, the EFFECTS of control actions are often either directly or indirectly observed (testing the product is, in essence, such an indirect observation). Our experience indicates, however, that the information ENTERING the controller is likely to be overlooked as an object of SPC treatment. In some systems, this is because such information is transient, i.e., it is used momentarily by the control software, and then discarded. And in any event, if these parameters vary in some other manner than they're supposed to, the controller will fix it, right? RIGHT! Heuristically, it's precisely BECAUSE the control scheme is designed to keep certain parameters on target, regardless of system status, taht these parameters are not likely to e good indicators of what is going on behind the scenes. As argued above, the results of control actions will generally allow detection of process problems only AFTER they have progressed to the point of inducing some significant system failure--and then it's too late. Controller input data, on the other hand, will typically be the most current information available on key process parameters (that's why it's sent to the control scheme, after all). Monitoring it at this juncture maximizes the quality system's potential to detect incipient process excursions before their severity becomes a serious challenge to the control systems' capacity to respond.

Although experience to date suggests that statistical monitoring inside the control "loop" has merit in most automated control environments, it is particularly so in situations where the control model is an empirical one (the chemicals industry is a prime example). Such models have limited domains of application; detecting deviations from normal process conditions before these bounds are exceeded is therefore an even more pressing need than in applications where the validity of the control scheme is known a priori (as in machining or satellite tracking, to name two disparate examples).

Model Performance Data. The adequacy of the controller's process model, or, more precisely, how its adequacy may be confirmed, its the source of yet another immensely valuable type of process monitoring data. Clearly this is an issue crucial to the proper function of an automated system. After all, having accurate, real-time data on process status and high technology process hardware is of little value if, in critical situations, the controller (human or computerized) issues commands which make things worse!

To effectively deal with the potential for model inadequacy, one must somehow validate the effect of the control action.

Historically, this purpose has most often been served by taking measurements on the final product. The more complex the process, or the further upstream the action, the more indirect (and hence less precise) was this measurement for that function. (One might reasonably wonder whether the modern-day deemphasis of final product QC is an implicit recognition of this fact as it pertains to complex conventional manufacturing facilities.) For multi-step processes, properties of intermediate product stages fulfill the same information niche. In our view, it is in this capacity alone that final product monitoring (and, to a lesser extent, materials property monitoring in general) will continue to have value in the conduct of SQC (although, as remarked above, it may have other uses, such as product compliance verification).

Even this use may prove to be transitory, in the long run. Control schemes typically have the capability to perform, in addition to the necessary command functions, the optional task of predicting quantitatively their effects on key parameters being monitored (e.g., their own inputs). Comparison of subsequent performance to these predictions would provide DIRECT validation of controller function. Measures of agreement between actual and forecast values would be of great worth as objects of statistical process monitoring.[1] A major peripheral benefit of this approach would be validation of controller input data, errors in which would otherwise go undetected until product compliance checks indicated a nonconformance to specifications.

Other quantities of potential interest in this regard are the controller commands themselves (e. g., tool offsets, catalyst charges, valve settings, etc.). In many cases the controller's computations mirror the underlying fluctuations in the process itself, aiding an assessment of whether the model employed is satisfactory (as such, this data is useful for root cause analyses, as well as on-line monitoring).

---

[1] For obvious reasons, knowledgeable practitioners tend to favor measures of this sort which have known, simple, probabilistic properties (at least under the hypothesis of model adequacy). For example, the sequence of differences between the actual and forecasted values of a variable is often "white" (i.e., its successive values are uncorrelated, with constant variance and mean zero). Such a statistic is ideally suited to monitoring via standard control charting techniques (e.g., Shewhart or CUSUM charts).

Note that, since materials properties may be among the input parameters of interest, attention to such measurements is not likely to disappear from the SQC picture in the foreseeable future. Such materials variables would, however, cease to be an end in themselves, being relegated in an automated environment to a role commensurate with that of quantifiable parameters.

<u>Data on Process Function</u>. Suppose, finally, that a manufacturing process has functioned successfully thus far, that valid input data is reaching the controller, and that the control scheme is issuing the commands necessary to maintain nominal processing conditions. To assure correct plant function, it remains necessary that those commands be properly executed by the machine tools, reactors, assemblers, auxiliary hardware, etc., with which the process is equipped. Is the lathe operating at the correct rpm? Is the cutting insert at the proper location relative to its "home" position? Is the workpiece being held as it should be by the adaptor (or "fixture") which mates it to the chuck? Has a vessel been pressurized properly? Has a raw material recycle stream become contaminated? Are reactants being fed at the correct rates, and is catalyst being recovered at the predicted mole ratio? Are bolts being tightened with the specified torque? Has a manipulator properly positioned the correct subassembly?

All of these are items the status of which can be ascertained by measuring observable quantities. Such data is of considerable value for verifying process function, and, consequently, for early detection of deviations from acceptable performance. Two caveats should be noted here, however. First, many parameters of this sort will be the subject of direct compliance checks by local machine controller software. Such quantities will seldom be of value for process verification, since they would quickly be restored to nominal by a properly functioning controller (and could therefore be of use for validation monitoring, however). Second, one must be sure to avoid logical circularity in the control/monitoring cycle. Verifying the working pressure of a reactor with the same gauge which was used initially to set it, for example, would be a pointless act of self-deception.

Summary.   In   conclusion,   our   experience   with   emerging automated   systems   convinces   us   that,   in   a   CIM   environment, SPC   can   no   longer   viably   be   divorced   from   control.   As   a consequence,   candidate   variables   for   statistical   process monitoring   may   be   expected,   in   most   cases,   to   fall   into   one   or more   of   several   categories   relating   closely   to   the   plant's   control system.

Even   the   fairly   complicated   view   offered   above   is   somewhat oversimplified,   of   course,   since   most   real-world   systems   (as remarked   in   the   introduction)   will   consist   not   of   a   single controller,   but   of   a   distributed   network   of   automata.   As   a result,   the   identification   of   parameters   to   be   monitored   is   likely to   have   to   be   carried   out   on   several   partially   overlapping hierarchical   levels.   In   our   experience,   the   number   of reasonable   targets   for   this   treatment   can   easily   exceed   even that   feasible   in   an   automated   factory.   Prioritization   and   final selection   will   be   highly   location-specific   tasks   requiring   careful advance   planning.

The   Role   of   Statistical   Root   Cause   Analysis.   Unlike   their application   to   process   monitoring,   statistical   techniques   have not   been   widely   used   for   diagnosing   root   causes   in   the   quality systems   of   conventional   manufacturing   facilities.   To   be   sure, the   (necessarily   manual)   task   of   determining   the   reasons behind   process   excursions   has,   occasionally,   been   accomplished via   the   use   of   statistical   methods.   To   data,   however,   this approach   has   been   more   the   exception   than   the   rule.

We   see   the   rise   of   automated   manufacturing   as   adding   a   far stronger   analytical   component   to   the   practice   of   statistical quality   control   than   has   been   possible   in   the   past.   Properly exploited,   this   capability   has   the   potential   to   be   a   major contributing   factor   in   attaining   the   level   of   performance   sure   to be   demanded   of   the   factories   of   the   future.   Nevertheless,   the inertia   of   past   practice   has   led   statistical   diagnostics   to   be   the area   most   commonly   overlooked   by   developers   of   increasingly automated   quality   systems.

In   our   experience,   there   have   been   three   principal   reasons why   statistical   analyses   have   traditionally   played   such   a   limited diagnostic role in manufacturing. The first--inadequate computing resources--is already largely a thing of the past, and should certainly not be a factor in the facilities which are our subject here. The second, insufficient statistical expertise among factory personnel, remains an issue, but one which is easily addressed by hiring and/or training practices.

Rather, the limiting factor in recent years appears to have been a paucity of data (or, at least, of RELEVANT data ACCESSIBLE for analysis). Indeed, this fact is not surprising in conventional facilities. Data has often been manually gathered and recorded, or (perhaps worse still) manually kept in some areas of the plant, while being electronically collected and stored in others. Data collection is often prohibitively expensive, as was the electronic storage of large volumes of data prior to the relatively recent declines in the cost of electronic storage media. Hardcopy storage, on the other hand, is voluminous, usually incomplete, and seldom well organized. Furthermore, manual keypunching of hardcopy data for subsequent computerized analysis is so laborious and costly as to be infeasible under any but the most pressing of circumstances. Key process properties have therefore often gone unobserved as a matter of routine, or at least been inaccessible for timely use. The overall impact has been to cripple diagnostic analyses, which depend for their function on routinely kept process data (as opposed to data from stand-alone experimental programs). Significant gaps in the data render such undertakings impossible, or, at best, of unlikely success.

As remarked at the outset, this critical area of data acquisition and storage is one of those in which automated factories will differ most strikingly from conventional facilities. In view of the points raised above, the implications for the role of diagnostic statistical analyses are clear. Factory automation will make feasible for the first time the widespread use of statistical techniques for troubleshooting applications in the plant. For this potential to be realized, however, increased emphasis must be given to THOROUGHLY capturing the RIGHT data, and making it ACCESSIBLE via electronic means, in order to circumvent the informational barriers extant in conventional plants. That attainment of these goals can be FAR more readily achieved by effectively planning an automated facility, rather than by retrofitting a poorly conceived one, should be obvious, and CANNOT BE OVERSTATED.

One aspect of this task is to assure that any remaining manually generated data will be routinely stored on electronic media where it will be jointly accessible with information from automated sources. Significantly, the development of semi-automated instrumentation has lessened the labor-intensity, and hence the cost, of many data gathering operations which still require human involvement. Concurrent advances in commercially available database software, and in voice and optical data entry systems, offer the means to accomplish the once arduous conversion of such data to electronic storage. These and other strides in computer integrated manufacturing have made unified access to comprehensive data on plant operations a possibility even today. We may reasonable expect continued progress to lead to these total information systems becoming commonplace in the future.

Provided that personnel resources are in place, and that database integration issues are properly planned for, the factory of the future is therefore likely to be an environment uniquely suited to the application of statistical methods. This unprecedented potential to exploit the power of statistics has ramifications for fine-tuning process performance, assessing control strategies, understanding process economics, etc. From the quality standpoint, it affords plant staff and management the opportunity not only to detect developing problems sooner, but to investigate their root causes via modern analytical techniques, rather than through trial and error.

A Current-Generation System for Automated Plants. It should be clear form the foreground that, if the automated factory's capabilities for data capture and storage are exploited, then the magnitude of the task facing the statistical quality system will be such as to necessitate effective use of the plant's computing capacity for that purpose, as well.

We have outlined thus far the rationale for our contention that the quality systems of the future must, and will, be able to fulfill (at least) two essential functions: (1) the detection of deviations from established process norms; and, (2) the

identification of the cause(s) for such departures.[2]   We devote
the remainder of this section to an examination of the form a
current-generation system of this type might take using
existing statistical techniques, and analysis software
commercially available at present.

For purposes of automated monitoring, selected data must be
examined in a timely fashion after its entry into the plant
computer network in electronic form (either by having been
captured that way from the outset, or by having been converted
to an electronic medium subsequent to manual collection).   A
range of implementation schemes are possible here, depending
upon what constitutes "timely" review of the process under
study.   Data on each monitored variable could, for example, be
retained in a buffer or temporary file pending a batch status
update once every hour, day, etc.   Alternatively, each data item
could be individually reviewed on an essentially real-time basis
(i.e., to within system resource allocation constraints) at it
arrives in the network.   One can envision, as well, approaches
which lie between these extremes in terms of the immediacy of
treatment received by monitored data, or which expedite to
differing degrees the handling of the various data streams.
Although less expeditious updates will suffice in many contexts,
especially for the present, the need for the earliest possible
detection of aberrant performance will impel quality system
evolution in the direction of achieving as nearly immediate
response as possible to those variables deemed worthy of
statistical monitoring.   For this reason, and because of our own
experience with this line of development, we devote the
remainder of our attention to developing a schema for systems
of this type.

The issue of data handling is not, of course, quite as
straightforward as it may appear.   Not all data to be retained on

---

[2] The reader may well have remarked that in none of the foregoing has there been any indication of
an inherent need for the monitoring and analysis functions to be even conceptually, much less
operationally, distinct. With respect to detection and identification of off-nominal performance, in
particular, one can well envision techniques which would identify which of several factors was the
principal contributor to a problem, in the process of detecting the presence of the problem itself.
(Consider, by way of analogy, classical two-way analysis of variance, which estimates the effect
and determines the significance of each factor, along with its calculation of the significance of the
overall model.) We consider this an important area for research in automated quality systems, and
plan to convey some proposals and initial results in a subsequent paper.

the plant database will be earmarked for monitoring. Nor, for that matter, will all monitored data necessarily be retained in the database for subsequent use. Indeed, "the database" itself will likely not be a single entity, but (as remarked earlier) a distributed system resident on a number of distinct hardware units at various levels in the computing network hierarchy. It should be understood, then, that our remarks pertain to distributed processing of data as it becomes available to the appropriate processing node, and regardless of the data's source or ultimate physical destination.

Figure 2 illustrates schematically the configuration of a representative current-technology automated factory. The statistical quality software of such an installation will be part of a total information system which will also encompass database, control, scheduling, accounting, and many other software functions.

A statistical quality control system for such a plant, feasible now using existing technology, can be constructed along lines indicated by the schematic of Figure 3. Key characteristics of

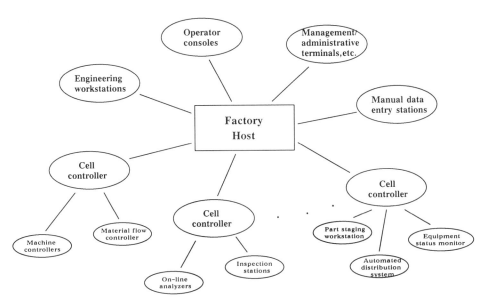

Figure 2. Plant Hardware Schematic

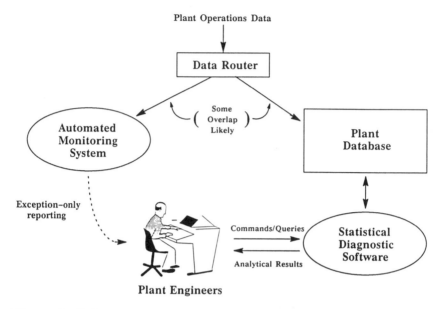

Figure 3. Software Schematic - Plant Quality System

such a prototypical system include: (1) automated on-line monitoring of selected data streams using standard control chart methodology; (2) automated assessment of charting results; (3) automatic reporting of out of control signals to plant engineers; (4) automated capture and long-term storage of data having value in diagnosing causes of process upsets; and (5) stand-alone general purpose statistical software for use in manually generated performance analyses. Points 1 and 2 are the principal areas where decisions concerning statistical methodology are required, and are the subject of Sections below. Nevertheless, a few general explanatory comments seem warranted at this point. First, note that the diagnostic and monitoring functions are operationally separate in this first-generation system, despite our remark in footnote 1. The reasons for this are: (1) the utter necessity (and technological feasibility) of automating the monitoring function, and (2) the current lack of sufficiently versatile and "intelligent" diagnostic systems. At present, therefore, we are constrained to mate user-run statistical analyses with computer-initiated monitoring routines. Fortunately, research in improved statistical techniques and in "smart" analysts' assistants will likely change this picture in the near future.   In the interim, the established vendor-supplied statistical software packages are the natural candidates for use in installing a manually directed diagnostic system.

Regardless of the method of implementation, however, certain essential characteristics of both the monitoring and diagnostic systems are self-evident.   The volume of data monitored clearly dictates, for example, the infeasibility of manual assessment even of computer-generated charts of the process--hence the need for an automated interpretive function, and for the exception-only involvement of human engineers once an out-of-control situation is detected.   Effective use of the diagnostic system requires, on the other hand, thorough information on the plant's operation.   Careful identification of data which will be of use in this regard, whether or not it has value for the monitoring function, it therefore a critical consideration during the planning process.

With such a system in place, plant engineering and management would be in a uniquely powerful position to observe, assess, and improve quality and productivity. Departures from accepted performance would be detected and reported as they occur.   Statistically trained process engineers could then respond as appropriate to identify and correct the problem, using statistical techniques as tools (perhaps assisted by a resident statistician, or by semi-expert advisory software, when needed).   With these capabilities at their disposal, the manufacturing staff of the near future will be equipped to aggressively pursue the goal of increasing efficiency while achieving near zero defect production.

Software Implementation of Statistical Monitoring.   In a fully computerized environment, the most natural implementation of control charting will not involve actual charts at all, of course. A control "chart" will consist, instead, of a subroutine with several associated parameter and data values stored in memory.   When new data arrives for charting, a call to this routine will use the new information to update and record the status of the "chart".   The stored parameters will retain auxiliary information (e.g., control limits) needed to assess whether the process is in control.   Stored data values (e.g. the last several values of the charted variable, or the value of a cumulative sum) will provide any historical information needed for this assessment, including the previous status of the chart. Note that, strictly speaking, each set of stored parameter and data values is analogous by itself to a control chart, since the same subroutine can be used to update all "charts" of a given type. The function of such a system is represented schematically in Figure 4.

It is important to remark for a moment upon the differences between such a system and most of the plethora of control charting software products now on the market. The mechanism described above is not a stand alone system, for example--it must interface with a data-gathering "front end", and with a "back end" designed to notify plant engineers when the process is out of control. More importantly, however, the scheme presented above ASSESSES and, if necessary, REPORTS, the chart's status, rather than acting as a graphics tool simply for DISPLAYING it. Moreover, such a system operates on-line,

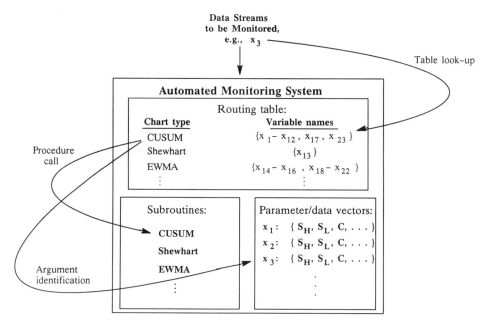

Figure 4. Schematic - Automated Control Charting

rather than being done retrospectively, as manual initiation of automated plotting usually is.

Software which has assessment capability need not, of course, be as complex or "intelligent" as it may at first appear. Control charting, throughout its history, has deliberately been developed to be as objective and algorithmic as possible, in order to facilitate rote application on the factory floor. It is therefore a technique uniquely suited to computer implementation; one merely writes the evaluation criteria into a computer program, rather than teaching them to operators or engineers. That most commercially available software does not routinely implement an array of out-of-control rules (or, for that matter, permit entry of data and parameters as arguments, rather than manually or from files) is an artifact of its intended use, rather than of any technological limitation.

Statistical Monitoring Methodology. The principal technological issue, from a statistician's viewpoint, is that of which of the generally accepted methods is best suited to implementation in this manner. Among general purpose statistical monitoring techniques, the Shewhart, CUSUM (cumulative sum), and EWMA (exponentially weighted moving average) control charts are the best known and most widely used. Of these, the Shewhart chart has long been valued for its arithmetical simplicity and intuitive graphical appeal. The CUSUM, on the other hand, it the more powerful statistically; this power, however, is purchased at the price of employing slightly more complex arithmetic, and of being considerably less intuitive to non-statisticians. The EWNA chart has performance characteristics similar to the CUSUM approach, while being slightly simpler to update, but having somewhat more computationally complex non-constant control limits. Its intuitive appeal is nearly comparable to that of the Shewhart school.

The performance deficit from which the Shewhart chart suffers can be largely offset by the addition of auxiliary rules (two of three points outside a set of warning limits, seven or eight successive points on the same side of the center line, etc.). Because of their purely enumerative nature, these rules, too, are simple for humans to apply. That they add logical complexity and require one to review the recent history of the process is of little concern to human analysts.

From these considerations it is evident why Shewhart charts have been favored for manual application in conventional facilities--and why CUSUM or EWMA charts should be for use in automated environments. The slightly more complex calculations associated with these more modern methods is of little consequence of a computer, and the Shewhart's greater intuitive appeal obviously of no benefit. Furthermore, the fact that modest process shifts and trends can be detected relatively quickly by CUSUM and EWMA charts, without resort to auxiliary rules, simplifies and shortens program logic required for software implementation. In addition, these charts minimize the storage volume required, since the only historical (or status) information needed is the most recent value of the cumulative sum(s) or moving average, respectively. Hence, the CUSUM and EWMA methodologies may be expected to consume less system resources in the completion of a given monitoring task.[3]

Our concern for efficient implementation may seem out of place in view of recent (and anticipated future) decreases in memory cost, and the computing power present in automated facilities. Yet our experience suggests that the variables of potential interest from a statistical monitoring standpoint may easily number in the thousands (dare we say tens of thousands?) in such installations. Unless management dedicates costly hardware specifically to the quality function (which seems unlikely in most instances unless we statisticians become considerably more popular in certain quarters than we are today), such a burden will likely be infeasible for the computing resources of even an automated factory to bear for some time to come. This is especially true in view of the fact that SPC will be but one of many plant functions being performed largely simultaneously ON-LINE. Thus, efficiency of implementation is likely to be a key factor in determining the thoroughness with which the plant's operation can be monitored within computing resource capacity constraints.

---

[3] An exception to this balance of preference may occur in cases where the use of multivariate (as opposed to several univariate) charts is considered essential. Multivariate generalizations of CUSUM, EWMA, and modified Shewhart charts have yet to be widely accepted, and, of the several forms which have been proposed, none appears to have yet gained dominance. The most common form of multivariate chart thus far is therefore the T-squared chart, a generalized form of the unmodified Shewhart methodology. Whether any current type of multivariate control chart offers sufficient practical advantage in most applications to warrant their considerable added computational and/or logical complexity, remains an object of spirited debate.

Summary. Powerful first-generation automated SPC systems can be built TODAY along the lines presented above. The development process benefits from being able to exploit existing SPC techniques and general purpose statistical software. When integrated with automated data collection and database management hardware and software, and placed at the disposal of a skilled engineering staff, such systems have unprecedented potential as quality and productivity tools. (Indeed, some systems of this sort are coming into operation even now, and have begun to demonstrate their usefulness.)

So far as we have seen at this writing, however, such systems are not yet available fully "off the shelf" (contrary, perhaps, to some vendors' claims). Software for automating the monitoring task (i.e., for routing appropriate data to a charting routine, and continuously updating and "interpreting" selected "charts") seems to be the only significant missing link. As commented earlier, the barrier here is not technological--it seems, rather, to be an issue of awareness, and of compatibility with the many hardware and software environments within which it may need to function. These problems are being addressed by several vendors at present, however, so this obstacle may soon disappear. In the interim, it is system component which may need to be custom engineered.

Some technical issues remain, as well. Although established control charting methods are feasible solutions to the monitoring problem, they cannot, we feel, be considered optimal. Moreover, the performance characteristics of systems comprised of many such charts has not been addressed until recently (Faltin, 1987; see also Woodall and Ncube, 1985) In the concluding section, we comment further on some of the work which remains to be done in these areas, as we speculate upon the nature of the next generation of automated SPC systems.

Conclusion: The Future of SPC in Automated Manufacturing

As is undoubtedly evident by now, we believe that, although its focus and implementation may change, the application of statistical process controls in automated manufacturing will be with us for some time to come. SPC's broadening interface with adaptive control is likely to be a key driving force in the future evolution of applied research in both fields; the fruitful advance of both must be in the direction of further integration, until they become indistinguishably intertwined.

As one might infer from our views regarding the synergy of these two disciplines, we feel strongly that the smoldering debate between the adherents of SPC, on the one hand, and those of control theory, on the other, is something of a red herring. Whatever the control strategy adopted in a particular setting, there will be a need for the foreseeable future to monitor, among other things, the function of the controller itself. Moreover, the numerous sources of variability in a manufacturing system place an in-principle limitation on the improvements realizable via optimal control. Reduction of these depends on achieving improved process knowledge - a task not explicitly treated by traditional control theory, but one to which statistical techniques for detection process changes are well suited. On the other hand, traditional SPC methods have, by their very emphasis on detection, tended to be too vague in their prescriptions on how to make needed process adjustments. In an age of complex processes under digital control, the algorithmic schemes offered by stochastic control theory could hardly be more timely.[4] Indeed, the growing use of diagnostic analyses in understanding and optimizing process functin can be expected to have the fine-tuning of control parameters as one of its most important missions.

We may expect this fusion of SPC and control theory to be more evident in the next generation of automated quality systems (it is not, really, in the scheme proposed in the preceding section). pertinent data on controller function and process model adequacy will be extracted by the control program itself, and reviewed by monitoring algorithms operating in parallel with the control routines. Implemented locally, the effect on overall system resources would be minimal. Exceptions, when they occur, would be reported not to

---

[4] Since the completion of the first draft of this paper, the integration of these two fields has emerged as an active area of applied research by a number of investigators, including the present authors. Our current nom de plume for the topic is Algorithmic Statistical Process Control, or ASPC.

plant engineers directly, but to a supervisory expert system which functions as part of a control "meta-loop". This supervisory system would be assigned the task of determining the appropriate response, e.g., an adjustment of control parameters elsewhere in the process, or a request for diagnostic analyses of recent data--a call for human intervention being a solution of last resort.[5]

The statistical algorithms which such systems can call upon should be no means be limited to those in use today. In the near term, of course, there is ample work to be done on improving upon existing methods, and better understanding their performance in system applications--important areas include further development of multivariate techniques, and methods for dealing with serially correlated observations (e.g., via time series).

In the longer term, the distinction between monitoring and diagnostic analyses is likely to become blurred. It would clearly be desirable to monitor the process in such a manner as to have some clue as to the cause(s) of any observed changes; moreover, the complexity of modern manufacturing processes invites such an approach. As a simple example, consider the multifactor nature of a machining facility in which a part may be cut on any of several machines, using any of several mounting fixtures. The monitoring approach adopted should ideally be one which identifies whether an out-of-control indication is due to one or more of the machines or fixtures, whether the problem is common to both factors, an interaction among them, etc. In other contexts, monitoring techniques which estimate the time of onset of an excursion might be used in conjunction with a log of process changes in order to suggest the identity of a casual variable.

Clearly, there is a wide range of issues awaiting research, and no lack of applications already at hand. If properly exploited,

---

[5] It is to be expected that future quality systems will be significantly be affected by advances in other fields, most notably, computer science. Expert systems, for example, may soon enter the QA systems picture not only in the form of expert supervisors, but as analysts' assistants in the conduct of diagnostic analyses, as well. Development of such expert statistical advisors is already well underway. Other lines of research (for example, natural language interfaces, and advanced vision systems) will also affect the implementation of statistical quality methods, even where they do not impact statistical methodology, per se.

these opportunities possess a genuine potential to profoundly alter the manufacturing environment of tomorrow. But even though the quality techniques of a few decades hence may be barely recognizable to those of us who practice today, they will be in many respects the lineal descendants of methods and philosophies we know well.

References

Faltin, F. W., (1988) "Approximate Run Lengths of Multi-Chart Control Systems", <u>Proceedings of Second National Symposium on Statistical Process Control</u>, Center for Professional Development, College of Engineering and Applied Sciences, Arizona State University, November 9-11, 1987.

Woodall, W. H. and Ncube M. M. (1985) "Multivatiate CUSUM Quality-Control Procedures", <u>Technometrics</u>, 27, 285-292.

INDEX